# SUSTAINABLE LAND MANAGEMENT
## IN THE TROPICS

# International Land Management Series

Series Editors: R.W. Dixon-Gough, Land Management Research Unit, University of East London, UK and Reinfried Mansberger, Universität für Bodenkultur Wien, Austria

This series is designed to encourage a study of the complex issues of land management in an international context and to emphasise the multi-disciplinary nature of land management. The key areas for the series will range widely from the concepts of environmental issues of sustainable land development, the acquisition of data, methodologies and a comparative analysis of current practices, the management of land information systems, land conservation, aspects concerning the regulation and legislation of land management, to human and social issues of land management.

*Also in the Series*

**The Role of the State and Individual in Sustainable Land Management**
Edited by Robert W. Dixon-Gough and Peter C. Bloch
ISBN 978-0-7546-3513-0

**Methodologies, Models and Instruments for Rural and Urban Land Management**
Edited by Mark Deakin, Robert Dixon-Gough and Reinfried Mansberger
ISBN 978-0-7546-3415-7

**The Management of Urban Development in Zambia**
Emmanuel Mutale
ISBN 978-0-7546-3596-3

# Sustainable Land Management in the Tropics

## Explaining the Miracle

*Edited by*

**KEES BURGER**

*Wageningen University, The Netherlands*

*and*

**FRED ZAAL**

*University of Amsterdam, The Netherlands*

Routledge
Taylor & Francis Group

LONDON AND NEW YORK

First published 2009 by Ashgate Publishing

2 Park Square, Milton Park, Abingdon, Oxon OX14 4RN
711 Third Avenue, New York, NY 10017, USA

*Routledge is an imprint of the Taylor & Francis Group, an informa business*

First issued in paperback 2016

**British Library Cataloguing in Publication Data**
Sustainable land management in the Tropics : explaining the
  miracle. - (International land management series)
  1. Sustainable agriculture - Africa, Sub-Saharan 2. Rural
  development - Africa, Sub-Saharan 3. Land use - Africa,
  Sub-Saharan 4. Sustainable agriculture - Tropics 5. Rural
  development - Tropics 6. Land use - Tropics
  I. Burger, Kees II. Zaal, F. (Fred)
  333.7'6'0967

**Library of Congress Cataloging-in-Publication Data**
Sustainable land management in the tropics : explaining the miracle / [edited] by Kees Burger and Fred Zaal.
    p. cm. -- (International land management series)
  Includes bibliographical references and index.
  ISBN 978-0-7546-4455-2 1. Land use--Economic aspects--Tropics. 2. Climatic changes--Economic aspects--Tropics. 3. Soil erosion--Tropics. 4. Sustainable development--Tropics. I. Burger, Kees. II. Zaal, F. (Fred)

  HD156.S87 2009
  333.730913--dc22

                                                                    2009000625


ISBN 978-0-7546-4455-2 (hbk)
ISBN 978-1-138-26681-0 (pbk)

# Contents

# List of Tables

# List of Figures

# Notes on Contributors

**Michael K. Bowen** is Lecturer at the Daystar University, Nairobi, Kenya. He did the research for this programme when at the Moi University, Eldoret, Kenya.

**Kees Burger** is Associate Professor with the Development Economics Group at Wageningen University. He was coordinator of this research programme when at the VU University Amsterdam, the Netherlands.

**Ton Dietz** is Professor of Human Geography at the University of Amsterdam, the Netherlands, and member of the Amsterdam Institute for Metropolitan and International Development Studies (AMIDSt).

**Esaïe Gandonou** is Lecturer at the University of Abomey-Calavi, Benin. He did the research for this programme when at the VU University, Amsterdam, the Netherlands.

**Wouter T. de Groot** is Professor at the Centre for Environmental Studies (CML), Leiden, and the Radboud University, Nijmegen, the Netherlands.

**Jan Willem Gunning** is Professor of Development Economics at the VU University Amsterdam, the Netherlands.

**Andries Klaasse Bos** was Lecturer with the Development Economics Group at the University of Amsterdam, the Netherlands.

**Henry K. Maritim** is Professor of Agricultural Economics and Dean in the School of Business and Economics, Moi University, Eldoret.

**Samuel M. Mwakubo** is Senior Lecturer at the School of Business and Economics, Moi University, Eldoret, Kenya, and a policy analyst at the Kenya Institute for Public Policy Research and Analysis (KIPPRA), Nairobi. He did the research for this programme when at the Moi University, Eldoret, Kenya.

**Remco H. Oostendorp** is Associate Professor of Development Economics at the VU University Amsterdam, the Netherlands.

**Marino R. Romero** is Professor at Isabela State University, Cabagan Campus, the Philippines. He did the research for this programme when at the Leiden University, the Netherlands.

**Wilson K. Yabann** is Associate Professor in the Department of Applied Environmental Social Sciences, School of Environmental Studies, Moi University. He is also the Dean of the School of Environmental Studies, Moi University, Eldoret, Kenya.

**Fred Zaal** is Senior Lecturer with the Amsterdam Institute for Metropolitan and International Development Studies (AMIDSt) at the University of Amsterdam, the Netherlands.

**Adri B. Zuiderwijk** was a post-doctoral researcher at the Centre for Environmental Studies (CML), Leiden University, the Netherlands.

# Preface

This book results from a research programme entitled 'The Agricultural Transition towards Sustainable Tropical Land Use', sponsored by the Netherlands' Science Foundation NWO, and particularly by its Environment and Economics Programme. The programme focused on the struggle against erosion fought by farmers in developing countries, and in particular by those farming on hillsides in semi-arid and sub-humid regions. In many places, the current agricultural practices were not considered sustainable in the sense of promising the same yields in the future, without a change in agricultural technology. This change in technology has already occurred in many places including some previously considered environmental disaster areas. The experiences in Machakos, Kenya, so appealingly documented in Tiffen, Mortimore and Gichuki's book *More People, Less Erosion*, triggered this research programme. The programme aimed at answering three research questions, formulated at that time (1998) as follows.

- Faced with a choice between a perceived future of continuing environmental degradation and a perceived future of environmental rehabilitation for which, however, investment must be made, what do and will farm households choose?
- What are the determinants of this decision: 'capacity factors' such as private endowments, knowledge and social capital; and 'motivational factors' of an economic, social and cultural nature, which in turn partially depend on environmental and contextual factors such as soil fertility vulnerability, roads to urban markets, rural population density and so on?
- How can these factors be translated into a valid and theoretically coherent economic model that describes the 'transition tendency' of a region, and is complemented by a sound insight into the non-modelled factors?

The formulation reflects the thinking of the time, in which the trade-off between present-day investments and future benefits played a prominent role. The research followed a microeconomic, actor-oriented approach and yet it ambitiously aimed at developing a regional 'transition-tendency indicator', that could more or less predict what a region, given its endowments and socioeconomic conditions, was likely to do: make a transition to sustainable land use or not. Among these endowments and conditions, room was made for factors beyond the standard economic variables, including social cohesion among farmers, learning abilities, interventions by civil and state organization and the like.

The programme was designed to investigate the micro actor-oriented foundations of the narrative of the development in Machakos and an obvious choice therefore was to collect farm household level data from surveys in that district. A natural extension then was made to also include in the research programme the somewhat poorer and drier neighbouring district of Kitui. These two districts should adequately cover the Kenyan situation; they provide data on farmers set in the same macroeconomic environment, but in different local settings. For a comparison with development in quite different macro settings, regions in Benin and Cameroon were chosen to represent areas in less developed conditions, and with poorer endowments in terms of nature, climate and income. The original idea was to develop the 'transition model' for these regions, and then apply this to humid tropical conditions, in this case the Philippines, to check its applicability to such new situation. This particular choice was also inspired by the existing links between universities, and working experiences of the initiators, that provided a common ground ranging from the terraces in the isolated Mandara mountains of northern Cameroon to the terraces established to provide Manila with vegetables.

The team consisted of three post-doctoral researchers, affiliated with the University of Amsterdam, the VU University Amsterdam and Leiden University; four PhD students, two in Kenya with the Moi University School of Environmental Studies, one in Benin's National University and one in the Philippines (at Isabela State University); and their supervisors in these countries and in the Netherlands. The team members had a diversity of academic backgrounds including economics, human geography, anthropology, environmental sciences and econometrics.

The programme was kicked off with a workshop held in Machakos, in which Mary Tiffen, Michael Mortimore and Francis Gichuki all participated. In her opening presentation, Mary Tiffen emphasized that time should be taken to delve into archives and literature to document the environmental and economic history of any region, and stressed how important oral histories were, as well as discussions with village leaders and experts. A series of parallel descriptions of change, as measured by different disciplinary techniques, was then re-examined to master the material and clarify the timing, sequences and interactions – and identify cause and effects.

The approach emerging from the workshop was to start with a selection of four villages per region that differed in terms of population density and distance to a major market. Of these villages, rather detailed descriptions were made, with special attention to the history of activities in relation to soil and water conservation. In each village some 25 households were randomly selected. A detailed questionnaire was developed, mostly similar across the 20 villages, with questions on the crops grown in all plots of the farm household, the characteristics of the plots and the households, the inputs used for crop and any soil and water conservation, and the market activities of households, including sales, purchases and off-farm activities. As cross-section information allows only limited interpretation of causal relationships, a few questions were included on the timing of soil and water conservation; in Kenya a second round of data collection was held in the year 2000.

By 1999, two years after the Machakos workshop, a new workshop was organized to evaluate the data and other information thus collected. While this proved a bit too early in view of the time needed for processing and cleaning the data, a joint 'feel' for what the data could show resulted from this workshop. The first impression was one of tremendous diversity, both between the villages and between the households. Distance to the market came out as more important for investments in land than population pressure; the costs of some of these investments were surprisingly low: stone bunds and soil and grass constructions of (sloping) terraces are not particularly costly, but often require annual reconstruction. Stone and rock terraces are expensive, however, but sturdy.

The four PhD students continued their work after this workshop and finalized their theses some years later.

This volume presents the insights gained in this research programme. The next chapter investigates the responses to work done by Tiffen et al. by looking at the many reviews that were made of their book. The following chapters, 2 and 3, focus on the choice of level at which the 'transition' can be understood. The data collected at regional, village, household and plot level enable such an analysis. Chapter 2 contains a statistical analysis of the Kenyan data first on regional level variables that play such a prominent role in the study by Tiffen et al. The question whether other factors contributed to the adoption of technological innovations that improved productivity and conservation was answered in the affirmative: macro-conditions matter.

Yet, when walking the hills and fields of Machakos and Kitui, it is clear that very many farms and fields remain unterraced or unprotected, and that these lie next to neighbouring farms and fields that are very carefully tilled and cared for indeed. How much of the variance can be explained at the level of households or plots? Chapter 3 shows that both levels are indeed relevant to the discussion on processes of innovation in agricultural systems. There is interaction between higher- and lower-level variables within the process of technological change in agricultural development. The households typically differ in their degree of market integration, and – connected with this – their resilience to shocks. Chapter 4 investigates their importance empirically for the case of Kenya by looking at the impacts of market access and risk. Access is operationalized in terms of time needed to find a buyer, and in terms of distance. Distance to market and distance of the plot to the house have negative effects on soil conservation, and so does risk.

Benin and Cameroon provide a setting in which farm households are at larger distance from the market. The groups in the two countries both have a history of erosion control, as local population densities have in the past made land scarce. The population pressure has eased in both regions, however, as northern Cameroon has become more peaceful, allowing a massive migration from mountains to plains, and northern Benin found ways to send migrants to other parts of Benin. Based on Beninese data, Chapter 5 provides detailed econometric estimates of the positive effects of soil and water conservation; it shows that such effects are only found when one looks within households, rather than between households. This underscores the specificity of each farm household. Chapter 6 looks at the impact that distance has on the position of individual households and shows that more remote households act as if labour is more plentiful and food more scarce. This seems not to favour soil conservation, however. Chapter 7 confirms this for the case of the Koza plains in northern Cameroon. While households did have knowledge of soil conservation as a consequence of the time when they were confined to the mountains, little conservation is done now that the plains can be peacefully cultivated. Lack of organic matter and other fertility-enhancing inputs appear to prevent productivity from increasing. The final case study, in the Philippines, concerns an entirely different area as to climate and many other variables. Yet, as Chapter 8 shows, the importance of markets in promoting the terracing of hillsides is again confirmed. Lower-level variables are also important, and this case seems to be the only region studied where the government plays a prominent role.

In Chapter 9 a first conclusion is drawn in that the theory of induced innovation, which is a form of Boserupian development, is applied to the cases at hand. Where soil and water conservation devices are not particularly expensive, the choice is for a technology rather than for a capital good. Such technological changes quite normally result from changes in the relative scarcities of land and labour. Where land is relatively scarce, land quality is enhanced. In Chapter 10 a synthesis of the findings is provided with a view to further conclusions and policy implications. The theoretical arguments are taken up again and compared with the findings of the case studies. The questions at the start of the programme relating to household trajectories and transition indicators appear to have faded in favour of a richer view in which households embrace opportunities offered by the market, and adjust their land quality only in relation to its relative scarcity and when productivity can be drastically enhanced. But scarcity of a location-bound factor such as land is only felt locally, and only those who experience its scarcity and its value switch to a land-augmenting, labour-intensive technology. Sustainability is the byproduct.

The Editors

# Chapter 1

# Optimistic Determinism or Explaining a Miracle

Ton Dietz with Jan Willem Gunning, Andries Klaasse Bos
and Adri Zuiderwijk

## Introduction

A long tradition in the study of sustainable rural and agricultural development has culminated in the book *More People, Less Erosion*, which tried to establish in a concrete situation whether this process of induced change can actually be proven to have happened. It is a case study (of Machakos District) in a particular period (sixty years, between the 1930s and 1990s) and that means it is only a case study, albeit a very extensive one and studied from a very wide angle. However, the timing of the publication and the changed paradigm that spoke from its pages soon helped to bring attention to this work, and it remains one of very few examples of a thorough study of the relationship between population growth, technological development and the standards of living in developing regions. In this chapter, we aim to reflect on the publication by Tiffen, Mortimore and Gichuki to assess its impact, and to discuss the reviews that were written upon the publication of the book.

## A Landmark in Development Studies Theory

*More People, Less Erosion* quickly received a 'star status' in geographical, environmental and development circles, and a little in the domain of agricultural economics. It was written by Mary Tiffen, Michael Mortimore, both from England, and their Kenyan co-author Francis Gichuki, based on research carried out in 1990–91. The book was published in 1993/1994, by John Wiley & Sons in England, and simultaneously in Kenya by ACTS Press. The study was the result of a research project funded and carried out by the Overseas Development Institute in London, which also published some of the preliminary results as working papers. Within a few years the 1993/1994 book was reviewed by many relevant journals, and by many of the leading authors in the field, particularly those from Britain (see Annex A1 to this chapter). It was picked up very fast by *The Economist* (with a review in December 1993), and by *The Independent* (in June 1994). In the academic domain of 'planning and development' the book was reviewed by *Public Administration and Development* (by Shepherd), by the *European Journal of Development Research* (by Lund) and by the *Development Policy Review* (by Upton) in 1994; by the *Journal of Development Studies* (by Clayton) in 1995 and by the *Third World Planning Review* (by Sage) in 1996. In the discipline of 'environmental studies' *Land Degradation and Rehabilitation* was first to publish a review of the book (by an anonymous reviewer) in 1994, followed by the *Journal of Arid Environments* (by Thomas, also in 1994), the *International Journal of Environmental Studies* (by Brown) in 1995 and by *Disasters* (by Downing) in 1996. In the domain of 'geography', the *Transactions of the Institute of British Geographers* took the lead (by Briggs) in 1995, and one of the leading British geographers, Gould, singled out a summary chapter of the book which was included in a compilation *People and Environment in Africa* (Mortimore and Tiffen 1996, in Binns 1996) as a summary of a 'mold-breaking book' in his otherwise rather critical review of Binns's book in the same edition of *Transactions* (Gould 1996). Another prominent British geographer, Adams, followed with a review in *The Geographical Journal* in the same year, and there even was a review in a journal for physical geographers (*Earth Surface Processes and Land Reforms* by Richards, also in 1996). In the discipline of agricultural studies and agricultural economics Macarthur reviewed the book in the *Journal of Agricultural Economics*, and Parton in the *Australian Journal of Agricultural Economics*, both in 1994. Some more practitioners' journals followed soon (*African Farming*, the *ILEIA Newsletter* and *Pesticides News*). The influential journal *Agricultural Systems* also carried a review, in 1996 (by Ssali). Finally, in the domain of 'African Studies' the French *Cahiers d'Etudes Africaines* was quick to publish a review in 1994 (by Thébaud), introducing the book in the French-speaking world, including French-speaking West Africa. The *Journal of Southern African Studies* followed suit (review by McGregor). The prestigious *Bulletin of the School of Oriental and African Studies of the University of London* included a review as well (by Allan, in 1995), and *African Affairs* followed in 1996 (with a review by Kenworthy).

The book was also widely cited in scientific journals. The ISI citation index[1] mentions 318 references in ISI journals to the book between 1994 and 2007. Google scholar[2] gives 771 hits and Google itself gives 17,700 hits for a search on 'more people less erosion tiffen', 11,900 if 'Mortimore' is added to the search terms and 9620 if the third author, Gichuki, is added as well.[3] Around the time of publishing the book, Mary Tiffen and Michael Mortimore also published a scientific article in *Outlook on Agriculture* (Tiffen and Mortimore 1993, December), which was more or less ignored (with only one citation in an ISI journal afterwards). A journal publication in *World Development*, one of the leading journals in development studies (Tiffen and Mortimore 1994), was much more successful though, receiving

---

1　　http://isiknowledge.com; this includes any inaccurate references to the book, e.g. citing the wrong years of publication (searched on 19 November 2007).

2　　http://scholar.google.nl/ (searched on 19 November 2007).

3　　http://www.google.nl/ (searched on 19 November 2007).

25 citations in ISI journals afterwards.[4] In addition there were two scientific articles in *Environment* (Mortimore and Tiffen 1994, 1995), which have received respectively nine and one ISI citations, and a scientific article in *Development and Change* (Tiffen 1995), which attracted five ISI citations. Before 1993 some preliminary work had been published as well (e.g. Tiffen 1991), but 1993–95 really saw an avalanche of publication activities around the 'Machakos story', and the framing of that story with a theoretical 'Malthus defeated by Boserup' line of reasoning. In 1996 a summary of the book was published in a compilation of 'environment and population in Africa' articles for university educational purposes (Mortimore and Tiffen 1996, in Binns 1996). Binns's book was cited 16 times, with special attention for the chapter on Machakos.

The success of the publications approach can be illustrated by showing the citation history of the book and the most important journal articles (see Figure 1.1, based on Annex A1 to this chapter). The book was clearly much more successful in drawing the attention of scholars than were the three journals, although the journals seem to be well chosen as leading journals in the domain of development studies and environmental studies. In total the 1993–95 publications were cited at least 350 times, and the secondary citations reached more than 2800. The various publications were cited most in the year 1999, and authors who cited Tiffen et al. in 1999 (and in 2001) were also cited a lot afterwards. The period around the launching of the Millennium Development Goals was also the peak period of using the 'Machakos miracle' as a positive counterpoint to the doom scenarios for Africa's predicament.

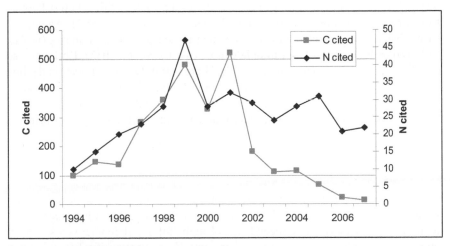

**Figure 1.1**    **Citation history of 1993–95 publications by Tiffen et al. according to ISI**

---

4    Google Scholar gives 37 citations.

Six publications which referred to the Machakos case were particularly influential. Frank Ellis's article on household strategies and rural livelihood diversification in the *Journal of Development Studies* (Ellis 1998) introduced the book to a wider field of development specialists and livelihood researchers. Ian Scoones engaged in a debate about new ecology and the social sciences in the *Annual Review of Anthropology* (Scoones 1999) and hence brought the book to the attention of anthropologists. Jesse Ribot used ideas from the book in an article on Africa (Ribot 1999), which introduced the ideas to Africanist circles in general and social forestry and public administration scholars in particular. Philippe le Billon and colleagues used the book in a study about natural resources and armed conflicts in *Political Geography* (le Billon 2001), and hence further introduced the book to scholars in the domain of politics and geography. Finally Tiffin and Mortimore's first (1994) article in *Environment* got a major facelift when Helmut Geist and Eric Lambin referred to it in a very influential contribution in *Global Environmental Change-Human and Policy Dimensions* in 2001 (Geist and Lambin 2001). This journal is widely read by climate change researchers.

The book inspired authors dealing with all parts of the developing world. The 350 references to the book and the journal articles in Figure 1.1 were used in 101 journal articles with general titles, and 257 dealing with a particular country or (sub)continent. Not surprisingly journal articles about Africa dominated the regional publications (n=218), and if the titles were more specific, most were dealing with East and North-East Africa (n=76), followed by West Africa (n=50) and Southern Africa (n=31). But authors also dealing with Asia (n=29) and the Americas (n=10) referred to this Machakos story as a source of inspiration, or – in a few cases – criticism (one group of authors deliberately used a counter-title: 'Fewer People, Less Erosion', in a publication about Bolivia; Preston et al. 1997). Again, not surprising, Kenya dominated the journal attention for particular countries (n=41), followed by Nigeria (n=11), Ethiopia/Eritrea (n=10), South Africa, Burkina Faso and Tanzania (all n=8) and Uganda (n=7). There was only one publication dealing with environmental problems in Europe (about Italy), which referred to the book, and one dealing with environmental problems in North America (about Canada). The Machakos story clearly appealed to the community of development-oriented researchers, but not so much to the 'mainstream sciences'.

*Praise and criticism*

Our chapter aims, first, to summarize a considerable number of book reviews on some selected key topics which are relevant for the subject and have been dealt with by most reviewers, and, second, to look at the follow-up in later years. Consequently the chapter may not give a representative summary of what individual reviewers have written, but we shall try to do justice to everyone by making direct quotations from each review. Due to the selection of critical items, together with a certain selection of critical and controversial statements, our chapter may suggest that reviewers are more critical than they in fact are. We therefore want to start off by noticing that the attitude of all reviewers without exception is sympathetic and positive. Any author would be pleased to receive attention from so many and such stimulating critics from the various disciplines as apparently represented by the twenty-odd referees. Just a sample of relevant statements:

> This book makes exciting reading. It clearly demonstrates that medium and low-potential rain-fed areas can support rapid population growth and high population densities with improved soil and water conservation. (Upton 1994, p. 328)

> The true significance of this book is not that it challenges established orthodoxies in a way that happens to be politically convenient, it is that someone got down to the task of asking sensible and open questions about change over a useful time period (i.e. decades) in semi-arid Africa. Machakos is unusual in having such a rich history of previous studies, but even here this synthesis was not previously available. Such research is time consuming, but it is vital if we are to break out of the endless cycle of simplistic blueprint 'answers' for Africa. (Adams 1996)

> This book will take its place in the halls of imperfect resource theory (a crowded shelf) and a prominent place in historical landscape ecology. On the whole it is a widely cited book that few will find the patience to read carefully and critique. It should be an essential read for researchers of Kenyan development policy and its manifestation in sub-humid and semi-arid areas. (Downing 1996, p. 91)

The publications by Tiffen et al. have been praised for their thorough analysis and fascinating research results, which apparently have surprised many reviewers because of the positive and encouraging outcomes with respect to the combination of high population growth, maintaining income levels per capita and sustainable land use. Terms such as 'Machakos miracle' and 'development paradox' have been used frequently, indicating that many researchers tend to think in negative scenarios with respect to economic growth and environmental consequences of sustained high population growth rates. This could be expected, given the overall negative overtones of the bulk of earlier scientific publications on the just mentioned 'triangle' of population pressure – standard of living – land use.

More than twenty book reviews have been brought together in this chapter and we have listed them in Annex A2. In this chapter we 'shorthand' the reference to the reviews by mentioning each author's name; and the reader can trace the corresponding journal in the annex. In seven sections we will summarise the main elements of the book reviews and add our own critical questions and interpretations.

## Causality of the Relationship between Population Pressure and Innovation

In the conceptual model presented by Tiffen et al. high population growth over a long time period leads to high population pressure and this will start a number of changes including technological development and farm investments, which all together result in the higher productivity of land and even labour, so that an increased income per capita situation will be the end result. It is the proof of Esther Boserup's theory (Boserup 1965, 1981). The authors even extend it towards environmental aspects to arrive at sustainable economic development. As can be learnt from their model the starting point is population pressure. 'Authors single out population density as the main driving force and use it to explain why it took decades for the environmental recovery process in Machakos to start' (Ssali, p. 325). 'Do more people mean less erosion? Yes at least in this place-time example, but not necessarily as a causal relationship' (Downing, p. 91).

The book's chapter on technical change shows that particularly after 1950 the process of technical change, in its demonstrated forms, was accompanied by a dramatic population growth rate. Both phenomena are closely associated, but cause–effect relationships are only given in a suggestive way, not in a quantitative presentation. 'Testing causality between population and environment requires more formal quantitative modelling and comparative case studies' (Downing, p. 91). 'The present study is replete with description, but totally lacking in this sort of modelling or statistical analysis and with only the rudiments of sensitivity testing' (*ibid.*). Downing refers to the methodological dilemma posed in the last chapter of Tiffen et al. under the heading 'Population Policies': 'The Machakos experience between 1930 and 1990 lends no support to the view that population growth, even rapid population growth, leads inexorably to environmental degradation. It is impossible to show that a reduced rate of population growth might have had a more beneficial effect on the environment' (Tiffen et al. 1994, p. 284). Downing states that such questions could be solved if testing of the relationships had been undertaken. We are more hesitant. The model presented by Tiffen et al. is already rather complex, and even then one could question whether all relevant factors have been included. Is it possible to develop the conceptual model as a quantifiable model, given this complexity?

In his review of the book, Clayton is more careful compared to Downing as regards the relationship between population growth and environment and uses the word 'compatible'. 'The authors provide convincing evidence that population increase, fivefold in the period, is compatible with environmental recovery [ ... ]. They rightly observe that a critical ingredient for this to occur is the availability of markets for profitable farming. In 60 years the value of output per hectare has increased tenfold and the value of output per capita approximately threefold' (Clayton, p. 638). But he continues on the relationships as follows. 'The authors posit, on the lines of Boserup, that the growth of population in the district [ ... ] is an important *causal* [italics ours] factor in generating new market opportunities which have stimulated investment and innovation – though vital concomitants have been the initiative and enterprise of the Akamba people themselves, with support, rather than top-down intrusion, from government' (ibid.). So various new elements are stressed by Clayton which could explain any income increase for farm households as resulting from their own initiative and decision-making on the basis of new market incentives and their own technological know-how, but also with some support (not intrusion) from the government. Interrelationships become complicated and one could question the causality of high population growth in the whole system. Similarly one could question whether the complex network of relations between variables, as portrayed in the book (Figure 16.1) resulting in higher per capita incomes, must start so one-directionally at stage one: population growth. For instance: what is the role of external economic factors like new market opportunities, or on the contrary economic contraction of the national economy?

## Population Pressure and Land–Labour Relationships

In the review contributions, various comments are made on the land–population relations as land–labour relations. In Chapter 4 of the book, population figures show the high population growth rates both in Kenya and Machakos. Based on census figures, the population of Machakos over the 60-year period has grown sixfold (1932: 239,000; 1989: 1.4 million inhabitants), although Table 4.1 in the book shows that the growth rate of Machakos has been lower than the Kenyan average except for during the period 1969–79. One should be a bit careful with these growth rates, though, as the book is. In the book (Tiffin et al. 1994, p. 62) it is mentioned that the sixfold increase might be an exaggeration. Population growth has no doubt been considerable but may not have been as high as stated: the 1932 census has most probably been an underestimate, while the 1989 census might have been 'inflated' (the accuracy of that census is questionable). What is not mentioned, though, is the fact that post-Independence censuses have always been organized in August, a time in the year when many children and young adults studying elsewhere returned for their holidays and when many men working elsewhere came 'home'. Those people were then counted in their 'home area'/area of origin, instead of in their area of work/study, where they were residing during most of the year. The actual population and labour availability during most of the year is considerably lower than indicated during census time.

The logical conclusion seems to be: 'The expanding population has increasingly placed pressure on the land [...] The reduced size of holdings has led to a typical pattern of intensification in the farming system' (Tuley). But the potentially negative influence of high population growth on rural household incomes is not as dramatic as could be expected, since total available land increased considerably after 1962 with former 'Crown lands' being made available, the percentage cropped land that continuously increased (due to a changed land use pattern away from livestock grazing) and a growing part of the male population being employed outside agriculture. Interestingly, from 1932 up to 1979 the cropped acreage per agricultural worker *increased* from 0.5 to 1.05; and even the cropped acreage per person has not fallen (Tiffen et al. 1994, Table 4.6). These few statistics just illustrate that the quantitative impact of high population growth on the man/land relationship is not that dramatic and explain why reviewers do not pay much attention to poverty due to increased population pressure over time. Much more dramatic is the qualitative aspect that, as demonstrated by various contrasting pictures, landscapes were already seriously deteriorated at the start of the 60-year study period and under the usual paradigm environmental recovery could not be expected. However, the authors prove, again with convincing photographs, that the environment had recovered as a result of activities of thousands of farmers (with women being major contributors), and this has been the main reason for the 'Machakos miracle'.

We note, though, that it is surprising that few reviewers carefully differentiate 'increasing population density' from 'increasing population pressure'. The book itself is also not very clear about the definition of 'population pressure' (it features prominently in Figure 2.5 on p. 28, but is not defined in the list of definitions on p. 29 and also not included in the index). Population pressure is not only the shrinking capacity to feed a growing population with locally available food resources. The external market may play a considerable intermediary role, where locally produced goods are being exchanged for externally produced food items at (potentially) positive terms of trade. Also a lot of 'pressure' may be relieved if the local economy is being supplemented by external funds: labour remittances, food and other aid, free or heavily subsidized goods and services. Also, population figures are usually referring to districts or parts of districts. At the same time, we know that this area, though relatively densely populated, certainly had land available for expansion. This refers us to the problem of scale dependency of the analysis (see further discussion below).

### The Role of Technological Change

The essence of the Machakos story is that particularly after 1960 many farmers applied technological improvements that resulted in large increases over time in land productivity. What is important to say from the outset is that the book authors carefully avoid the impression of a big bang set of innovations, a kind of 'Green Revolution', changing the agricultural and environmental conditions more or less overnight 'continued incremental, adaptive changes, many of which are hardly noticed ... can add up to substantial change in the aggregate' (foreword in Tiffen et al. 1994, p. 7).

Five important agricultural-technological innovations were improved maize production (particularly the flexible incorporation of the short-cycle 'Katumani' maize variety into the cropping system, and the adoption of double cropping), the introduction of horticultural crops, fruit trees and coffee, the ox-plough, the use of compost and manure, and a trend towards stall-feeding and fodder growing, as well as tree planting and bench terrace construction.

Technological changes were aimed at increased productivity per cultivated land and labour unit. The role of farmers themselves in the application of changed technologies is pre-eminent. The book reviews accept the importance of these agro-technical changes but do not give them any specific attention, unlike the other important breakthrough: the development of sustainable agro-ecological practices (soil and water conservation mainly by means of terracing).

> The authors demonstrate that population growth was instrumental in preventing erosion and environmental degradation. (*Pesticides News*)

> The authors are exploring a case which contradicts much of the general and simplistic pessimism prevailing on the discussion on natural resources management in Sub-Saharan Africa. The myths that population increase inevitably leads to land degradation [...] and that land degradation generally is irreversible, are undermined by the Machakos reality presented in this study. (Lund, p. 195)

The study teaches us that the main ingredients for realizing – over a long period – improved and sustainable agriculture are in soil conservation. The special chapter dealing with developments in soil conservation divides the conservation history of Machakos into four periods: 1930–45; 1946–62; 1962–78 and 1978–90. From the viewpoint of massive changes by farmers in erosion control the period of 1962–78 is most interesting. By 1961 the area conserved by two types of terraces, the most prominent feature in soil and water conservation in the area, had fallen to 27,000 ha, compared with a peak of 42,000 ha in 1958. The cultivated area was about 110,000 ha (Tiffen et al. 1994, p.194). During the short rains of 1961 much damage was done by abnormally heavy rainfall. Officially, compulsion (very strong in years before) was ruled out in the period around Kenya's Independence in 1963. Part of the anti-colonial atmosphere of the 1950s can even be attributed to the harsh environmental policies during that period. Observers in the early 1960s regarded the relaxation of 'environmental law and order' as having potentially devastating effects. Closed grazing areas were reopened and red 'sores', the forerunners of serious erosion, began to reappear in 1962–64. At the same time agricultural staff numbers were cut back. Grazing controls, soil and water conservation and controlled settlement largely ceased. About that same period various initiatives of farmers have been mentioned in the book, illustrating how they paid serious attention to soil conservation, despite government 'withdrawal'.

**The Level of Analysis and Individual Rationality**

The state of soil conservation in the early 1990s shows clear improvements, quantitatively as well as qualitatively, in terracing and other soil conservation elements. In an economic exercise comparing costs and benefits, at farm level, of soil conservation practice (compared with farming without soil conservation) it proves economically profitable to do so (Tiffen et al. 1994, p. 200).

In the book reviews, the environmental recovery has been noticed and appreciated. It is seen as primarily the result of farmers' decision-making. However,

> The book is ostensibly concerned with development of farming families, but the reader is ultimately given little idea how households secure their livelihood needs or how they make decisions about allocation of resources [...] Although they dot the landscape of the photographs, the Akamba are not accorded a 'voice'. The lack of personal testimonies in the book left me feeling suspicious that the principal architects of 'conservation' were not given the opportunity to express their experiences: perhaps their understanding of environmental change is at variance with that of outsiders who see only evidence of 'recovery'? (Sage 1996, p. 264)

Other reviewers have accepted the positive facts of environmental recovery, but various critics considered the process by which changes have arrived as obscure:

> These phenomena and events correlate or concur with the process of environmental recovery in Machakos, but we are left somewhat in the dark as to why this is so. Self-help groups, Christian missions, education and expansion of cash crop production are not restricted to the [sic] Machakos but can be found in many areas, which fared less well. [...] Machakos evidently is a very dynamic and adaptive society. We are, however, not brought to an understanding of the dynamics from the *actors' perspective* [our italics]. (Lund, p. 196)

There seems to be a lack of sufficient 'grounded' causal reasoning and a chapter is missing bridging the empirical parts and the theoretical sixteenth chapter in the book. Some critical comments by others also refer to lack of insight in the process and the weight of each factor involved.

In a contribution to a workshop in 1998 to prepare for a follow-up research of a Dutch–African–Asian research team (the start of the process that led to this book), where Mary Tiffen, Michael Mortimore and Francis Gichuki were present, Aad Zuiderwijk criticized the approach taken by Tiffen et al. for producing 'much circumstantial evidence, but with few eye witnesses': 'what lacks are eye-witnesses; the people who made the investments, and who can tell us a lot on what they did, when they did it, why, and how [...] No major effort was put to interview sufficient [numbers of] farmers in different socio-economic positions and agro-ecological zones' (Zuiderwijk 1998). The inclusion of the life histories of a few people gives the impression of a people-centred book, but it is not such a book and certainly not in a systematic way. As a result, we do not get an idea about the downside of agricultural intensification. Who are the losers? What about the socioeconomic (and sociocultural) differentiation in the area? What about changing relationships within communities and with the outside world? Income diversification and the diversification of the regional economy (with a lot of growth in transport, trade and real estate development, and important contributions from remittances) are major driving forces of investments in agricultural intensification, so it seems, but who does and who does not?

*Economics vs. anthropology: the integration of disciplines*

Economists and economic anthropologists would have loved to see more calculations and more life histories in the book, showing how investments in agriculture and in terraces could have been so rewarding, that it was indeed worthwhile for diversifying farmers to do so. And what was the historical order? Did investments in terraces (and in agricultural technology in general) follow periods of high rewards per area and per labour hour? Or did investments in terraces result in higher rewards per area and per labour hour?

This brings us back to the 'farmers did it' story. The downplaying of government agencies as drivers of change by Tiffen et al. in their conclusions, and the highlighting of farmers' own initiatives as a response to market forces (which many reviewers have also picked up as a major element of the study), is not always convincing. About the more recent terracing activities in the late 1980s and early 1990s Tiffen et al. write (on pp. 200–201): 'food-for-work and tools-for-work have helped poorer farmers achieve terraces through *mwethya* groups [so-called self-help groups, which were often assisted by government and non-governmental agencies], but hired labour has been used by those with the necessary resources'. So, relatively rich farmers can do it alone; the others need external support and encouragement? And one can even go one step further: did the farmers who 'did it alone' actually do it alone? Isn't it more realistic to say that they used a lot of cheap, hired, local labour, which had become available in the area due to the fact that so many poor farmers did not benefit from market changes and land improvements, as they only had minimal land areas, and did not benefit from marketing of crops, as they had few anyway, and certainly not the crops with occasional windfall profits?

In a 1995 study of the same Machakos area, a group of authors from a political ecology background put more emphasis on the historical political economy of the area; the differentiation between relatively rich, successful, and self-reliant farmers and a considerable group of poor, impoverishing households. They write:

> For over a century, Ukambani, the home of the Akamba people, has been the
> object of intense scrutiny and repeated interventions by international and national

'experts'. Outsider narratives have portrayed the region as a crucible for a series of crises, including human and livestock epidemics, 'overgrazing', soil erosion, low productivity, underdevelopment, fuelwood shortage, biodiversity loss, and threatened wildlife. Akamba farmers and herders recount a very different story in which land alienation, land hunger, and limits on mobility of people and their herds have restructured the ecological and spatial order of their homeland, to the benefit of some and the detriment of many. The history of crisis construction and resolution by outsiders, juxtaposed with the diverse experience of people within the region suggests that simple solutions to single problems may actually create new crisis, in Ukambani and elsewhere. (Rocheleau et al. 1995, p.1037)

### *Replicability and path dependency: local actors, local conditions*

However, how 'special' is Machakos? Both for scientific understanding and for development practice a crucial question is how a transition towards sustainability can be induced on a larger scale and what the conditions favouring such processes of change would be? Responding to this requires a detailed understanding of the factors that induce farmers to invest in farming systems that are sustainable. And it also calls for proper (and not ideologically motivated) analysis of the role of government agencies in some phases, and with enough attention for geographical differentiation: it might well be that in some areas farmers can be the main driving forces of terracing and other investments in environmentally sustainable agriculture, but that in other areas they need an external lead agent (in most of Machakos the government played that role during the last decades of the colonial era) provoking change, despised as it often was, and that in still other areas farmers will not be able to invest, neither now nor in the foreseeable future. If that geographical specificity is needed in Machakos, and we think it is, the questions of where, when and by whom beg for more theoretical attention.

Many reviewers see the 'Machakos miracle' as a good example of sustainable management of land use in a fragile environment. Tiffen et al. (1994) also put their story in this perspective: it is a book about 'the replacement of natural vegetation by sustainable farming systems, which over time maintain an adequate level of nutrient replacement, and which conserve soil and water in forms useful to man' (Tiffen et al. 1994, p. 14). However, further analysis asks for a careful and clear breakdown and operationalization of the concept of 'sustainable management of land use'. Here the book already gives a lead in its down-to-earth definition (p. 29): 'the maintenance or improvement, over several years (of fluctuating rainfall), of soil chemical and physical properties on cultivated land, of pasture productivity on grazing land, of farm trees and regenerative woodland communities, and of groundwater recharge, compared with conditions at a chosen baseline (or the commencement of a period of study or observation)'.

The authors of the book summarize their findings about sustainability on pp. 242 and 261–2. On soil chemical properties (soil fertility levels) they write that 'they have been unable to reach firm conclusions', although agricultural output per hectare has increased considerably and that would have been very difficult with declining soil fertility levels; on the other hand all farmers complained about problems of obtaining sufficient manure from their animals and of finding the cash for purchasing fertilizers; with a decreasing grazing land/crop land ratio – in the early 1990s 1.5:1 – this may become a major bottleneck in the nutrient cycling system. On soil physical properties they write that 'soil erosion has been eliminated on much cultivated land, and greatly reduced on others'. On soil texture there is 'a trend towards more sand, at the expense of the silt and clay fractions'. On pasture productivity they write: 'there are beginning to be signs of improvements in grazing lands'. On trees they write that 'the fuel shortage [...] has never reached the often predicted crisis point, and there are now more trees, grown for many different purposes'; and there is no conclusion about groundwater recharge.

For any follow-up comparative research it is important to use the same definitions, operationalization and measurement approach. What is also crucial is the *chosen baseline*. In the book the chapter on rainfall has mainly been used to show a rather extreme variability and unpredictability but especially the fact that there has not been a trend in rainfall. If there would have been a positive rainfall trend, this could at least partly have explained the higher agricultural yields and the vegetation coverage. However, more attention could have been given in the book to the impact of bad years (droughts, but excess rainfall or diseases/pests can also cause major problems) on changes in land and crop management during and immediately after such bad years.

### Replicability and Path Dependency: The Geography and History of the Machakos Case

Where history and geography meet, there is always the question about the adequacy of 'time slices' and 'area cuts'. This is not a topic many reviewers take seriously. We do.

Some presentation of evidence is done at the level of the Akamba area as a whole (so including Kitui); most presentation of evidence takes the (old) district as a spatial level of scale (with the problem that before Independence the Machakos Reserve was different from post-Independence Machakos District), and finally there is a presentation of important evidence at a lower level of scale.

There is a lot of suggestive explanation in the book where – due to paucity of data – the writers take whatever exemplary sub-district cases are available (e.g. Nzaui on p. 157, or Yatta on p. 172) and they add their own in-depth study locations (see the book's map on p. 4). However, at this level of in-depth study areas the 'weights of evidence' differ: looking at the number of times case-study evidence is being presented, most attention was given to Masii (an area with 51–100 inh/ km$^2$ in 1932 and 1948 and 100–200 in 1962 and 1979, see p.49). This is followed by Kangundo (26–50 in 1932, 100–200 in 1948, 200–400 in 1962 and 400+ in

1979) and by Makueni (less than 25 in 1932, 1948, and probably also 1962, and 50–100 in 1979. The other areas which are presented as 'study locations' get less attention: Mbiuni, Mbooni, Ngwata and Kalama. How representative are these specific areas for trends in the whole district? And if the study areas differ so much in the crucial variables (population density and population growth) wouldn't it be useful for modelling purposes to differentiate them according to a typology, e.g. a typology of land pressure (if that is possible)? Also, the distance to Nairobi and the role of coffee production should both have been given more attention. Are the most convincing pieces of evidence in coffee areas, near Nairobi? If so, what are the gradients to lower sustainability and less successful innovations away from the coffee zones, and away from Nairobi? Wouldn't it be true that economic processes of market-related intensification would be far more important than population pressure as such if this geographical aspect would be taken into account and could explain much of the process of achieving sustainability in dryland agriculture?

On the 'time slices' we can conclude that, where the book presents 'hard evidence' there is a remarkable emphasis on the early 1960s, and the late 1970s, hardly anything on the 1950s, late 1960s and early 1970s, and relatively little on the more recent period. It would be interesting to discuss if this is important or not with regard to the conclusions that are reached. What is intriguing, though, is the relationship suggested by Tiffen et al. (1994, p. 88) between terracing and 'increased market demand, from Kenyan towns and from export markets, transmitted by private traders'. This market demand particularly focuses on coffee, fruit and vegetable production, while much of the growth of that market-led expansion (re)started in 1974, accelerated in 1976–79 for coffee, and became relevant for fruit and vegetables mainly from 1980 onwards. At the same time, terracing had already started in the 1940s, had become very widespread in 1978, and mostly preceded the market boom (Tiffen et al. 1994, pp. 69–71). Reading the book, one often wonders: what happened when, where and in what order, and more systematically collected detailed life and investment histories would have helped to solve that riddle. We have tried to do that in this book (see Chapters 2 and 3).

The book by Tiffen et al. ends with a chapter called 'Replicability, Sustainability and Policy.' The question 'unique or replicable' states a number of factors that make Machakos rather unique, while other factors can be added as well. The authors believe that the differences with other areas are in most cases relative rather than substantive. This includes colonial land occupation and the subsequent availability of new land at Independence, suddenly relieving the tight man/land ratio. In our view this makes Kenya, or at least Machakos, a special case. However, the research team tends to be carefully optimistic:

> Comparative reviews of farming systems show that increasing population density correlates with crop-livestock integration, as well as with intensification, in all the major ecological zones of tropical Africa [...]. The growth of the non-farm sector [such an important explanatory factor in Machakos] is also common

[…]. Such comparative studies indicate that the Machakos experience is being replicated elsewhere and is likely to have wide applicability. (Tiffen et al. 1994, p. 276)

In the reviews various doubts and scepticisms have been put down with respect to replicability over time and place, though.

> Nor can it be assumed that proximity to the large urban market of Nairobi, and the relatively free markets for crops in Kenya, are not key factors in permitting an increase in the market surplus of agricultural produce from Machakos. In short, differences in current population density, quality of natural resources, location in relation to markets, and the general socio-economic environment in other parts of Africa might lead to very different results. (Upton, p. 329)

Also Ssali had his doubts:

> Machakos differs from other semi-arid areas in Africa in two ways: climate (bimodal rainfall and cooler temperatures); and unoccupied land (Crown land) that became available after Independence. (Ssali, p. 325)

Ssali believes that the influence of Nairobi, the mushrooming city next door to the Machakos District, may be underestimated in the book. They are challenging questions to answer: what makes Machakos a 'breakthrough case', what are the 'transition factors' to sustainable land use and how specific have they been? These questions are asked by follow-up research.

### The Follow-up: Towards More Comparative Analysis

After 1994–95 the book's success inspired other scientists to think about follow-up studies, designed to test some hypotheses and refine others. Our book is one of these, but Mary Tiffen and Michael Mortimore also designed and carried out a follow-up study, of which we will give a brief overview. But first we should highlight four other recent publications, in which 'Machakos' is put in perspective.

First, Steve Wiggins (2000) used some of the Machakos evidence in a comparative overview of 26 African cases (although he does not use the book but instead an earlier paper by Tiffen presented at a conference of economists in 1992; Tiffen 1992). He concludes that village studies show a rural Africa that gives less cause for alarm than the macro-level agricultural statistics from national agencies, which are mostly very worrying for the 1980s and 1990s. But he adds that the village-level studies all show that the crucial variable is market access.

Second, in a brief, but very illuminating contribution, Boyd and Slaymaker re-examined the hypothesis that population growth and agricultural intensification result in improved soil and water conservation, drawing on six new case studies from Burkina Faso, Ghana, Nigeria, Senegal, Tanzania and Uganda (Boyd and Slaymaker 2000). Their conclusions are rather sobering. They found hardly any other examples of a reversal of natural resource degradation and a trend towards environmental recovery. Environmental successes were limited to relatively small sections with high-value crops. Hence, soil and water conservation improvements will only be taken serious by farmers when these improvements have the potential to increase the yields of these high-value crops, when agricultural land is in short supply and when farmers still have a 'farm ethos'. Measures to support farmers to adopt land and farm improvements should be part of wider measures to support their overall livelihoods, which increase market access, and secure attractive producer prices.

Third, Murton revisited some of the Machakos evidence and came to conclusions that put far more emphasis on the losers in the process:

> changes in Machakos District, Kenya have been accompanied by a polarization of land holdings, differential trends in agricultural productivity, and a decline in food self sufficiency within the study area. [...] when the 'Machakos experience' of population growth and environmental transformation is examined at a household level, it is shown to be neither a homogenous experience nor a fully unproblematic one. (Murton 1999, p. 37)

Finally, Jules Siedenburg (2006) critically examined the Machakos evidence, and tried to put it in a balanced perspective, admiring the 'solid outcomes' but critiquing the 'unhelpful hyperbole' of the theoretical interpretations and much of the reception of the study. His comments:

> It is suggested that the Machakos study comprises hopeful data, on the one hand, and problematic calculations and assertions, on the other. After exploring problems with the study, the article suggests an alternative interpretation of the data that is arguably more pertinent to contemporary concerns with rural poverty and environmental degradation as well as more widely applicable in sub-Saharan Africa. (Siedenburg 2006, p. 75)

Let us see how Mary Tiffen and her team coped with these and other suggestions and criticisms when they designed and carried out a comparative follow-up study, which was funded by the Natural Resources Policy Research Programme of the UK Department for International Development (DfID). They took seriously the criticism that the 1994–95 studies were all so close to Nairobi that the urban influence might have been the main driving force and not increasing population densities. In Kenya they therefore did a study in a more remote and more arid part of Ukambani, the new Makueni District, towards the south. They also focused more than in the book on the 'policy requirements for farmer investments'. The new studies in Makueni were done with the major involvement of Francis Gichuki, the third author of the 1994 book, and working as a senior lecturer in soil and water engineering at the University of Nairobi, Department of Agricultural Engineering. They were mostly on water management (Gichuki 2000a–e). Studies were added on soil fertility, crop, livestock management and investments and income (Mbuvi 2000, Mbogoh 2000, Fall 2000, Nzioka 2000, Nelson 2000). Finally Francis Gichuki, Stephen Mbogoh, Mary Tiffen and Michael Mortimore produced a synthesis booklet (Gichuki et al. 2000). The studies show a design in which natural sciences and social sciences work alongside. The time depth is mainly between 1989 and 1998, which is a bit surprising, as the convincing power of the 1994 book had partly been based on the long time perspective.

The new study was intended to compare the Kenyan area with two other African dryland zones. The two other dryland areas which were added for in-depth analysis were Diourbel in Senegal and the Kano–Maradi area in Nigeria and Niger. A huge team of researchers participated in each of these studies. Twenty-four researchers worked on Diourbel, and together produced eleven working papers. The team was led by Abdou Fall of the Institut Sénégalais de Recherches Agricoles in Dakar (ISRA). A number of topics are the same as in Kenya: specific studies about water, soil and tree management, about crop and livestock development (during the 1960–99 period) and commercialization, about income diversification and farm investments, and about human resource elements (and particularly the functioning of institutions like the family and local support arrangements, and attention for education). In Senegal there was less specific analysis of rainfall trends, but more specific attention for demographic trends, the impact of national policies affecting farmers, land rights and access arrangements, and land use change and occupational change. In January 2001, a synthesis study was presented about the Diourbel Region (Faye et al. 2001). On Maradi in Niger and Kano in Nigeria comparable studies were done to the ones in Senegal, sometimes in joint working papers, often in specific documents for Maradi and Kano. For Maradi an English-language and a French-language synthesis was made (Mortimore et al. 2001a, 2001b), but none for the Kano area.

The proceedings of the concluding workshop (Drylands Research 2001) suggest agreement about four major issues: the importance of markets and of urban markets in particular, the importance of the rural non-farm sector, the importance of access to land, and the importance of local social institutions, in particular the institution of the family (and the way families manage their finances), the institution of education, and of values attached to education. However, quite a number of the critical points raised by the reviewers, and by us in our review of reviews, were more or less ignored by the participants of this workshop, and this is particularly true for the issues of social inequality, and for the impact of geography and distance to urban markets more specifically.

Instead of publishing a new book, Mary Tiffen and Michael Mortimore decided to use their new insights in a variety of journal articles, and to use their 'drylands website' for summarizing findings, and posting on-going work (see http://www.drylandsresearch.org.uk). Their focus was mainly on the Sahel (e.g. Mortimore 2001 and 2002), not so much on comparing Kenya and the West African cases (in fact only Tiffen 2002 did so). There was a lot of engagement with policy-making and thinking about the research–policy interface (e.g. Tiffen and Mortimore 2002; Mortimore 2003; especially Mortimore and Tiffen 2004). And this seems to be the major direction in which current work is going (e.g. Tiffen and Mortimore 2006). Still it is a pity that a real integration of the recent comparative study has not yet been published, and a comparison of these findings with the Machakos story also still needs to be done.

**Past Achievements and Future Work**

One publication of the follow-up project is becoming particularly influential in scholarly circles: the analysis of linkages between agricultural growth, urbanization and income growth in a publication in World Development (Tiffen and Mortimore 2003). It makes a strong plea for a major boost for urban productivity, in order to stimulate agricultural development and rural improvements. In fact the study acknowledges the importance of urban markets in any assessment of rural Africa's dynamics. However, one would then want to see how important distances to these urban markets are, how geography matters and what markets actually do. Despite a promising research design in the Makueni–Diourbel–Maradi–Kano comparison, systematic answers are still missing, though a recent article deals with the relationship between urbanization and agricultural change (Tiffen 2006). This then appears to be the overriding lesson from the 1994–2007 period: the prospects of rural environmental management and of agricultural change in Africa's rural areas depend on the development of urban demand, and instead of continuing with urban–rural divides in scholarly and policy circles, these domains should be combined for fruitful analysis.

**Annex A1**

| Year | Total | | Book | | World Developm. | | Environment | | Dev and change | |
|---|---|---|---|---|---|---|---|---|---|---|
| | N cited | C cited | N cited | C cited | N cited | C cited | N cited | C cited | N cited | C cited |
| 1994 | 10 | 97 | 9 | 93 | | | 1 | 4 | | |
| 1995 | 15 | 147 | 13 | 140 | | | 1 | 7 | 1 | 0 |
| 1996 | 20 | 136 | 19 | 135 | | | 1 | 1 | | |
| 1997 | 23 | 283 | 21 | 277 | | | 1 | 1 | 1 | 5 |
| 1998 | 28 | 359 | 25 | 301 | 1 | 5 | 2 | 53 | | |
| 1999 | 47 | 479 | 38 | 394 | 8 | 79 | 1 | 6 | | |
| 2000 | 28 | 326 | 25 | 263 | 1 | 40 | 1 | 5 | 1 | 18 |
| 2001 | 32 | 520 | 28 | 344 | 3 | 36 | 1 | 140 | | |
| 2002 | 29 | 181 | 26 | 178 | 3 | 3 | | | | |
| 2003 | 24 | 112 | 22 | 95 | 1 | 11 | 1 | 6 | | |
| 2004 | 28 | 115 | 24 | 81 | 4 | 34 | | | | |
| 2005 | 31 | 68 | 29 | 67 | 1 | 0 | | | 1 | 0 |
| 2006 | 21 | 21 | 19 | 21 | 1 | 0 | | | 1 | 0 |
| 2007 | 22 | 14 | 20 | 13 | 2 | 1 | | | | |
| Total | 358 | 2858 | 318 | 2402 | 25 | 209 | 10 | 223 | 5 | 23 |

**Annex A2**

| Name of reviewer | Name of journal, vol., no. and pp. | Year |
|---|---|---|
| W.M. Adams | *The Geographical Journal* 162, March (1), p.85 | 1996 |
| T. Allan | *Bulletin of the School of Oriental and African Studies* 58, p.430 | 1995 |
| N.N. | *African Farming* (Jan/Feb) | 1994 |
| J. Briggs | *Transactions of the Institute of British Geographers* 20 (4), pp.520–1 | 1995 |
| K. Brown | *International Journal of Environmental Studies* 49, pp.68–9 | 1995 |
| E. Clayton | *Journal of Development Studies* 31, April (4), pp.641–2 | 1995 |
| T.E. Downing | *Disasters* 20, March (1), pp.88–90 | 1996 |
| N.N. | *The Economist* (11 December), p.68 | 1993 |
| N.N. | *ILEIA Newsletter* (July) | 1994 |
| J.M. Kenworthy | *African Affairs* 95, April (379), 307–308 | 1996 |
| N.N. | *Land Degradation and Rehabilitation* (January) | 1994 |
| C. Lund | *European Journal of Development Research* 6 (2), pp.194–6 | 1994 |
| J. MacArthur | *Journal of Agricultural Economics* 45, September (3), pp.395–7 | 1994 |
| J. McGregor | *Journal of Southern African Studies* 20, June (2), 317–24 | 1994 |
| R. North | *The Independent* (20 June) | 1994 |
| N.N. | *Pesticides News* 23 (March) | 1994 |
| K. A. Parton | *Australian Journal of Agricultural Economics* 38, August (2), pp.208–10 | 1994 |
| K. Richards | *Earth Surface Processes and Land Reforms* 21 (8) | 1996 |
| C. Sage | *Geoscientist* 6 | 1995 |
| C. Sage | *Third World Planning Review* 18, May (2), pp.263–4 | 1996 |
| N.N. | *Spore* 49 (February), p.4 | 1994 |
| H. Ssali | *Agricultural Systems* 51 (1), pp.113–115 | 1996 |
| A. Shepherd | *Public Administration and Development* 14, August (3), p.317 | 1994 |
| B. Thébaud | *Cahiers d'Etudes Africaines* 34 | 1994 |
| D. Thomas | *Journal of Arid Environments* 28 (1), pp.82–3 | 1994 |
| C. Toulmin | *Africa* 65 (1), pp.152–3 | 1995 |
| S. Trumper | *Farm Africa Newsletter* (April) | 1994 |
| P. Tuley | *Tropical Agricultural Association Newsletter* (March) | 1994 |
| M. Upton | *Development Policy Review* 12, pp.328–34 | 1994 |
| W.S.K. Wasike | *The Environmentalist* | ?? |

Chapter 2

# Beyond Population Growth: Intensification and Conservation in Dryland Small-scale Agriculture; Machakos and Kitui Districts, Kenya[1]

Fred Zaal and Remco H. Oostendorp

## Introduction

The discussion on whether the agricultural population in dryland areas in Africa will follow a Malthusian 'poverty-trapped' or a Boserupean 'stepwise innovative' path has been raging for some time now.

The possibilities for transition of farming systems towards higher levels of productivity while still maintaining sustainability – defined here as the possibility for present generations to use the natural resources without compromising future levels of productivity – has been and continues to be a major concern of governments and international and multilateral organizations. Alarming messages abound of reduced availability of agricultural land and rapid and sustained population growth. Coupled with a continued reliance on agriculture these trends could endanger local agricultural societies and national food security. Declining or increasingly variable rainfall due to global climatic changes further threatens food production systems and food security at national level in many developing regions (Brown and Kane 1994; Van den Born et al. 1999; Dietz and Put 1999; Alexandratos 1999).

There is growing evidence that agricultural intensification, though by no means equivalent to increased sustainability of small-scale agricultural systems, can occur together with and contribute to it in a context of increasing pressure on lands (Boserup 1965; Conelly 1992; Tiffen et al. 1994; Reij et al. 1996b). Indigenous technology development and local testing and implementation of introduced technologies often achieve the limited goal of sustaining nutrient and organic matter contents in soils together with other goals of rural development (Richards et al. 1989; Reij and Waters-Bayer 2001). Institutional development and economic integration on the other hand may also have a positive impact, either directly or indirectly, on the motivation of and possibilities for farmers to invest in

---

1   This is a slightly revised version of Zaal and Oostendorp (2002).

the quality of their land and on the sustainability of management within the local land use and livelihood system.

In this chapter we aim to test the Boserupean hypothesis proposed by Tiffen et al. (1994) for the Machakos and Kitui districts by evaluating the role of other variables than population density in the process of intensification. We will be particularly looking at the *dynamics* of terrace adoption at the village level, making use of retrospective information on village-level variables such as population density, rainfall, crop prices (especially coffee) and terrace construction. The analysis will show that variables such as distance to major urban markets and windfall profits from the coffee boom in the late 1970s are at least as important for explaining the historical investments in the quality of land as increasing population pressure in the Machakos and Kitui districts.

The chapter is structured as follows. First we describe the study and study site on which the analysis of the determinants of agricultural intensification is based, including some of the relevant literature. In the next part we will describe the historical pattern of adoption for eight villages in Machakos and Kitui. Adoption was not smooth but involved a number of 'bursts' or 'peaks' during which villages went through rapid phases of agricultural intensification. We will describe these periods of heightened intensification activities, and we will relate them to certain events. We will present the results of a multivariate analysis in the next section of the chapter, where we estimate the determinants of agricultural intensification in Machakos and Kitui for the 30-year period 1966–95. We will also present a number of simulations to investigate the cumulative impact of the coffee boom at the end of the 1970s on terrace construction as well as the impact of the development of infrastructure, population density and droughts during this period. The final part follows with conclusions and policy recommendations regarding feasible approaches towards sustainable small-scale agricultural development in these dryland areas of Africa.

## The Study and the Study Site

The focus of research is on the context needed for farmers in drylands to not only increase productivity, but also improve the production environment. These terms are by no means equivalent.

Tiffen et al. (1994) focused on the role of population pressures, stressing the relationship between increasing population density and growing demand, labour availability, infrastructure and increased levels of interaction and innovation generation. They and others pointed at the evidence of actual Boserupian processes of population-growth-related innovation processes coupled with land quality enhancement (Templeton and Scherr 1997). However, other factors, such as market conditions, weather and government activities may be very important too (Brown and Shrestha 2000). Equally important may be local social conditions at village level, characteristics of households implementing the innovations (Lapar and Pandey 1999) and characteristics of the plots on which these innovations are applied (Pender 1999). All these levels need careful scrutiny before it can be concluded that population pressure is the main driving factor in practice.

A recent inventory of local soil and water conservation technologies showed that population densities do play a role as incentive to invest in land (as land becomes the scarcest resource, not labour). However, cases of low population densities with high levels of soil and water conservation adoption and cases where the opposite is found are too numerous to be able to say with confidence that it is this factor alone that causes the adoption of innovation (Reij et al. 1996b). For example, in Honduras, reverse trends and patchy occurrence of innovation and investments for conservation were found under conditions of increasing densities (Crowley and Carter 2000). Also, collective actions, considered by induced innovation theory to be related to high-density areas, were related to lower rather than higher density areas (Pender 1999). Of course, immediate benefits in the form of significant, recognizably and sustained higher yields from the innovations are very important, and may ultimately be the only incentive for farmers to adopt any innovation (Laman et al. 1996; Templeton and Scherr 1997; Zaal et al. 1998). Differences in adoption rates occur even within one area. Land fragmentation and unequal distribution may play a role here, making land more scarce for some people than for others. Gender plays an important role in this respect. Lack of credit facilities is often mentioned when lack of adoption of innovations is discussed, but it is not so much credit for these highly uncertain investments but rather lack of access to capital in general which is an important hindrance to investments in land. When money is available it may be invested in innovations, but money is not borrowed for this purpose. Regular remittances or windfall profits from high cash crop prices may therefore be important (Bevan et al. 1992, Bigsten 1996). More fundamentally, cultivation may be only one option in a larger portfolio of options (Ellis 2000). Other strategies for reaching sustainable livelihoods may be more interesting, for which other investments are needed. Thus, people may want to invest in the education of their children, the establishment of businesses, or livestock (Brons et al. 2000).

Innovation theory holds that the incentive to invest may be higher when the value of output increases, and for this either household demand needs to be high and few alternatives available, or the market prices should be interesting enough. At the same time transaction costs should be low enough to allow access to the market. A good price, input markets and credit availability, output marketing infrastructure and institutions (including information), institutions to manage resources, social organization in general and location in relation to markets are all-important for livelihoods based on natural resource use (Fleuret and Fleuret 1991; Templeton and Scherr 1997). In this sense the situation of small farmers in Africa is basically not different from that of any enterprise in the western world (Buch-Hansen 1992; Reij et al. 1996b). The point is: which innovations within intensifying agricultural development pathways are combined with investments in fertility enhancement, erosion control and agro-ecological diversity? The possibilities are often there but their realization depends on local conditions (Pender 1999; Conelly and Chaiken 2000). Finally, certainty in land rights may be a basic condition for sustained and high levels of investments in land-based innovations, if not the actual goals of these investments. These investments can be both soil-building and tenure-building (Gray and Kevane 2001).

Eight villages were included in the survey, with four villages in Machakos District and four villages in Kitui District. Research villages were selected on the basis of population density (from both densely and sparsely populated sub-location, the administrative level below the district and the lowest level for which data are available) and distance to Nairobi in travel time (both far and nearby, measured in minutes using public transport along the most direct road) as a proxy for transactions costs, so that consequently four categories were distinguished in each district. Table 2.1 presents the villages, the sub-location, and the scores of the various villages on the selection criteria. The scores are defined as A=(high density, low transaction costs), B=(low density, low transaction costs), C=(high density, high transaction costs), and D=(low density, high transaction costs).

Ecological conditions were kept constant as far as possible, by selecting villages in agro-ecological zone 4 (AEZ 4) (Jätzold and Schmidt 1982). AEZ 4 can be characterized as a transitional zone between semi-arid and semi-humid, depending on the altitude. It has between 115 and 145 growing days (medium to medium/short growing season) and mean annual temperatures between 15

**Table 2.1    Village characteristics**

| Village name | Sub-location | District | Sub-loc. Density (cap/km2) | Distance to Nairobi (min) | Category ** |
|---|---|---|---|---|---|
| Ngalalia | Ngiini | Machakos | 494 | 60 | A |
| Kisaki | Kithangaini | | 179 | 80 | B |
| Ngumo | Katheka | | 305 | 150 | C |
| Musoka | Kyamatula | | 121 | 145 | D |
| Range for rural Machakos | | | 30–1061* | 15–195 | |
| Range for AEZ4 in Machakos | | | 75–500 | 60–195 | |
| | | | | | |
| Mwanyani | Misewani | Kitui | 436 | 210 | A |
| Utwiini | Kaluva | | 64 | 195 | B |
| Kitungati/Matua | Kitungati | | 144 | 270 | C |
| Kyondoni | Kauwi | | 93 | 180 | D |
| Range for Kitui | | | 13–447 | 150–510 | |
| Range for AEZ4 in Kitui | | | 25–447 | 175–360 | |

*This excludes the two urban sub-locations of Mjini (1093) and Eastleigh (2825)
** A=(high density, low transaction costs), B=(low density, low transaction costs),
C=(high density, high transaction costs), and D=(low density, high transaction costs)

and 18°C in the Lower Highland Zone. The Upper Midland Zone has between 75 and 104 growing days (short to very short growing season) and mean annual temperatures between 21 and 14°C. Cattle and sheep keeping and the growing of barley are recommended in the Lower Highland Zone, while sunflower and maize are recommended in the Upper Midland Zone.

From each village, 25 households were randomly selected. This was done using a complete list of all households in the village, developed with the village elders and the village 'headman', the senior elder supposed to be the government representative at this lowest level. The final number of households visited depended on availability of these households and the possibility of finding replacement households for those households that were unwilling to answer the questions or that were not available. Table 2.2 gives the general information on the survey population size. All household members were enumerated. All plots, owned, rented out, rented in, in ownership or in use in any other way, were included in the survey and visited while the survey was implemented. GPS recordings were taken to be able to find the same plots in the second year of data collection in 2000.

**Table 2.2    General information on the survey population size**

|  | Number |
|---|---|
| Districts | 2 |
| Villages | 8 |
| Households | 193 |
| Household members | 1259 |
| Plots | 484 (422 valid on terracing) |

**The Dynamics of Agricultural Intensification**

*The level of explanation: village*

In this chapter we focus on the determinants of agricultural intensification at the level of the village, ignoring differences in adoption levels within the villages across households and plots. In principle this implies that much of the variance in adoption we leave unexplained, as actually most of the variation in the adoption of conservation techniques can be found at the household and plot level, as opposed to the village level. Here we are interested in village-level explanations of agricultural adoption for four reasons. First, much of the literature on agricultural intensification and the spread of innovations stresses explanations at this level of analysis, such as population density (Tiffen et al. 1994; Barbier and Bergeron 1998), transaction costs (Wadsworth and Swetnam 1998; Holloway et al. 2000), location and distance (Haegerstrand 1967), technological improvement (Barbier and Bergeron 1998), social structure (Havens 1975) and crop prices (Barbier and

Barberon 1998). Second, village-level (and higher-level) analyses are often most relevant for policy-making purposes, as most policies of agricultural intensification are policies of regional development. Third, we do not have retrospective data on household and plot characteristics, making a study of household- and plot-level determinants of adoption over time unfeasible. Fourth, adoption at the plot-level may well be affected by (endogenous) village-wide adoption patterns because of copying effects, technological spillovers and endogenous village prices (Pomp and Burger 1995; Taylor and Adelman 1996). By analysing reduced form patterns at the village level we avoid modelling such interactions, which are difficult to handle in household- or plot-level analyses.

We are aware that our analysis may be biased because of omitted household and plot characteristics in the analysis. For this reason, we will also compare our results with those based on a model of adoption at the *plot level*, instead of at the village level. Although we do not have information on time-varying household and plot characteristics, we will include time-invariant household and plot characteristics to test for the importance of household and plot heterogeneity.

In the remainder of this section we will describe the history of soil conservation in Machakos and Kitui districts over the past forty years. The analysis here will be descriptive and will focus on the trends in soil conservation activities over the entire period, as well as periods of 'peaks' or 'bursts' in soil conservation investment activities. Most of the intensification took place in these peak years, and therefore we will also look at a number of variables which may have played a role here, particularly variations in rainfall, increases in population density, implementations of agricultural development programmes, new road construction and variation in coffee and maize prices over the past forty years. This descriptive analysis provides the background for a more formal multivariate analysis later in the chapter estimating the impact of each of these factors on soil conservation activity.

*Forty years of soil conservation in Machakos and Kitui: trends, peaks and external factors*

Terracing overwhelmingly features as the most prominent type of investment in land quality. Terraces in this area are of the 'Fanya Juu' type, where trenches are dug along the contour and the soil thrown uphill, so forming the start of a terrace. Of the 422 plots for which we have valid data, 318 fields were terraced and 104 were not. For this reason, because of the resources involved and because of the role of terracing in maintaining moisture, nutrients and organic matter in the soils, terracing was chosen as the indicator of investment in land for both intensification and sustainability.

The adoption process in Machakos and Kitui districts is presented in Figures 2.1 and 2.2, with the fourth order polynomial trend line added in Figure 2.1.

In the two districts taken together (shown in Figure 2.1), after a slow start, the adoption of terraces speeded up until some years ago, when growth slowed down. This slowing down of the adoption process has been caused by a reduction of new

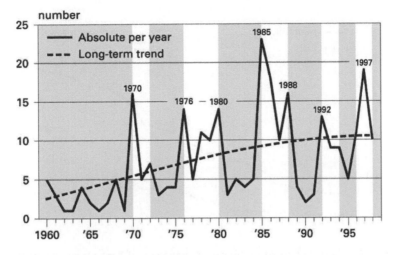

**Figure 2.1    Number of plots terraced, per year of first terracing on the plot, absolute for both districts, 1960–98**

terracing in Machakos, where most plots suitable for terracing have been treated. Kitui is still in the rapid adoption phase. In the case of Machakos, the total number of plots terraced was 214, with 40 plots remaining (15 per cent). In Kitui, 104 plots were terraced with 64 plots remaining (38 per cent).

The general trend is that of adoption of terracing on most plots in the districts. This trend may be related to higher-level variables such as population growth and growth in population density. The population density figures are as presented in Figure 2.3.

The adoption of terracing seems to follow the increasing population density. However, comparing Figures 2.2 and 2.3, we note that the population density of Kitui is still lower in 1998 than that of Machakos in 1960, while terracing in Kitui in 1998 is much higher than terracing in Machakos in 1960. This suggests that population density is not the sole factor in terracing. Also, in Figure 2.1, the occurrence of peaks suggests that other factors are at work as well and these peaks may be linked to certain events.[2] Five of the identifiable peaks are selected. In chronological order these are:[3]

---

2    The accuracy of linking the peaks in terrace adoption with certain events depends on the accuracy of the memories of the respondents, a notoriously unreliable source of information generally. In this case however the problem may not be the actual year so much as the fact that the memories of people will link the terracing with the year the event occurred instead of with the year after when they reacted to the event by starting terracing. The year people said they acquired the plots does not have a tendency to be linked to five- and ten-year periods.

3    Earlier peaks are much smaller and very difficult to relate to any events due to a lack of precise data. Generally, this was a period in which terracing was prescribed by the

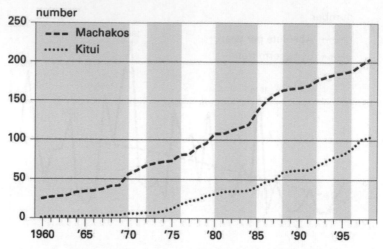

**Figure 2.2    Number of plots terraced, per year of first terracing of the plot in Machakos and Kitui Districts, cumulative, 1960–98**

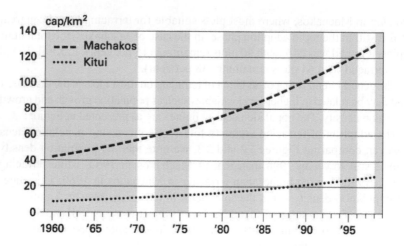

**Figure 2.3    Development of population density in Machakos and Kitui Districts (cap/km²), 1960–98**

*Note*: the population density figures are from the 1989 and 1963 census, as well as various District Development Plans (DDP). Interim years have been calculated using the growth percentages as presented in the DDPs.

1. Rapid adoption in 1970–72.
2. Rapid adoption in 1976 and again in 1978–80.
3. Very rapid adoption in 1985 until 1988.
4. Rapid adoption in 1992, somewhat less in 1993–94.
5. Rapid adoption in 1996–98, with a peak in 1997.

Chronologically, the first peak occurred in 1970–72, with three subsequent years of slightly lower adoption figures. This peak followed after a year of relatively low rainfall (see Figure 2.4), though no critical conditions seem to have been experienced as far as seriously affected crops and food shortages were concerned. As rainfall reappears again as a factor during a later period of adoption (see below), it therefore is indicated as one of the variables for inclusion in the model. Rainfall figures (average per year for the two districts, five-year moving average and fourth order polynomial trend line) are presented in Figure 2.4.

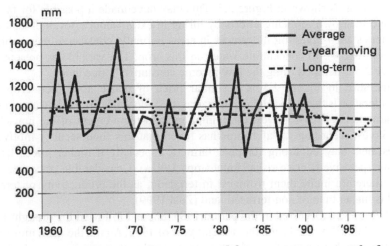

**Figure 2.4    Rainfall figures, long-term and five-year average trends, for both districts, 1960–98**

A more appreciable and sustained period of adoption of terracing is found in the period 1976–80. In fact, there are two peaks of which one may be the direct reaction to drought conditions in the period 1972–76 with rainfall figures around normal in 1974, but with shortages of between 20 and 35 per cent in the other years. There may have been an urgent need for soil moisture control.

During this time however terracing was also stimulated by the Machakos Integrated Development Programme (MIDP), which was set up in 1978 after

colonial government and implemented using various approaches including forced labour (IFAD 1992).

a long period of absence of any coordinated effort to initiate development on a programme or project basis.[4] MIDP lasted until 1988 when it ran out of funds and was followed up by the so-called Arid and Semi-Arid Lands (ASAL) Programmes in Makueni. However, only six per cent of the MIDP budget of 17.25 million Kenyan pounds for both the first period and the next one of 1989–91 (as an ASAL programme) was spent on conservation activities directly. In Kitui, no similar programme existed to supplement the efforts of the local population until 1982, when a USAID-funded ASAL programme was set up in the district. This lasted until 1997, with the Danish Development Agency DANIDA having taken up funding after 1990. In itself however, the presence of development programmes and projects is not a very useful independent variable. Much depends on the actual activities and interventions. Road construction as far as it was funded by MIDP (1978–82) may have been important for example (Tiffen et al. 1994).

Probably more important than the financial support of MIDP and ASAL was that the local population itself started to invest in terracing, following the rapid rise of world market coffee prices.[5] Prices of this cash crop soared between 1972 and 1978, as is shown in Figure 2.5. This may have made it possible for farmers to fund their own terracing efforts by hiring in extra labour (Bevan et al. 1992). Farmers started investing in terraces in areas normally considered unsuitable for coffee (AEZ 4), to concentrate water on this crop. The variable as presented in Figure 2.5 is deflated using the low-income consumers' price index.

Included in Figure 2.5 is a similar price of maize as a food crop. The consideration is that a price hike of maize as the main staple crop may cause interest of farmers to at least harvest the minimum food requirement of the household, and invest accordingly. Fafchamps has noted that this food security motive for investment may be strong for poor farmers, especially if higher prices lead to higher (perceived) price risks (Fafchamps 1992a). Considerable price increases may therefore incite them to invest in terracing, as the productivity of terraced land is higher than of non-terraced land (Zaal 1999).

A third peak occurred in the period 1985–88, and followed the drought in the early 1980s (Figure 2.4) and the drought year of 1987. Again, the main motivation to adopt may have come from moisture control rather than erosion as such, though the National Soil and Water Conservation (SWC) Programme, supported by SIDA from 1978 onwards, may have played a role as well. Though the SWC Programme started in the high potential areas (AEZ 2 and 3), it extended its activities to the

---

4    The last effort was the African Land Development fund (ALDEV), which started as a colonial land development programme in 1946 and lasted until 1962. Initially, the attention was mostly on land conservation and grazing control in African areas severely eroded; after 1951 the goals were more broadly defined (Tiffen et al. 1994, p. 254).

5    It is estimated for Machakos that the ALDEV contribution formed about 35 per cent of total investment in terracing until 1985, while private investment from 1985 onwards added 15 per cent to the total invested sum per year with little new contribution from project or government sources (Tiffen et al. 1994, p. 259).

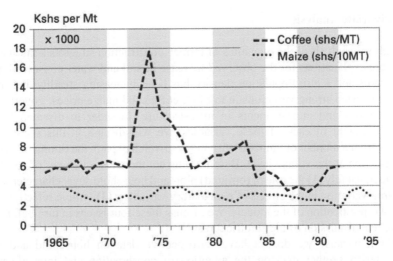

**Figure 2.5**    **National coffee and maize prices in Kshs per Mt of green coffee and white maize (indexed at 1964 price levels), 1964–95**

drier areas later on. However, from our evidence it did not appear to have been active in the villages in our study. In addition, this period witnessed the construction of the Machakos–Kitui road (Tiffen et al. 1994). Transport possibilities between these places improved considerably because of this and this may have stimulated market gardening and terrace building.

The fourth peak of the early 1990s followed the bad rainfall situation in 1990 and 1991, particularly in Kitui District. By this time, there was no longer a Machakos District development project to support terracing,[6] but in Kitui the earlier-established Kitui Integrated Development Programme (KIDP) had been renamed Kitui Agricultural Project (KAP) and had started to focus on agriculture exclusively. Still, with diminishing funds both through KAP and the Ministry of Agriculture and Livestock Development, the effects of external interventions was probably minimal. This peak and the last one in the second half of the 1990s may again have occurred in reaction to low rainfall figures and the need for better moisture control, not a generally felt need for erosion control per se. Also, maize prices rose by unprecedented percentages. Most of the terraces in these two last peaks were built in Kitui as by this time the majority of the plots in Machakos District had already been terraced.

---

6    Nor was it very much needed for this purpose any longer. The Machakos ASAL project had been terminated in 1991, while in the early 1990s the KIDP was in turmoil due to changes in the donor agency responsible and many other internal developments.

**Multivariate Analysis**

The above analysis of the dynamics of terrace construction suggests a number of possible explanations for the observed increase in soil and water conservation in Machakos and Kitui. Population density has been increasing steadily, but there have also been strong variations in rainfall, coffee and maize prices; government interventions and improvements in infrastructure. In order to disentangle the relative weight of each of these variables, we will discuss in this section the results from a number of multivariate analyses, which explain the timing of terrace construction at the village level. Before doing so, there are a number of issues to consider, namely (1) the role of (omitted) plot- and household-level characteristics, (2) time-variation in the effects of explanatory variables, (3) sample selection bias and (4) specification of the models. We discuss these four issues in turn before we move to a discussion of the empirical results.

First, because we do not have retrospective data on household and plot characteristics other than on the adoption of conservation and intensification measures, our analysis focuses primarily (but not exclusively) on explanatory factors at the village level. This implies a possible *omitted variable bias*, if the omitted plot and household level variables are correlated with the village-level explanatory variables. For instance, if villages that are located further from the market also have poorer households, and if poorer households are less likely to invest because of imperfect credit markets, then in our analysis the effect of distance to market on investment will also pick up this wealth effect. To control for this we also include in our analysis village dummies reflecting village differences in household and plot characteristics. Still this may not be sufficient if the relevant village-level household and plot characteristics vary over time, for instance because plots which are easier to be terraced are terraced first or because wealthier households are more likely to terrace earlier. This would imply that the relevant village-level household and plot characteristics will change over time, as the remaining plots may be the most difficult to terrace or owned by households who are least inclined to invest. We will test for the importance of such changes in the relevant household and plot characteristics by comparing the results of our village-level analyses with a plot-level analysis, where we allow for household and plot heterogeneity. The results show that our conclusions are unaffected in the sense that the same village-level variables explain plot-level adoption, although the results are less robust to the inclusion of year effects.

The second issue is the time-variant effect of the explanatory variables, as the number of terraces is reduced with each new terrace constructed. The various explanatory variables may have a different weight on terracing on the remaining plots, and as the plots that are easier to terrace are the first, each new plot terraced represents a greater effort as well. In our analysis we will allow for this possibility by testing for the presence of *structural breaks* between different time periods in the model.

Third, our analysis may suffer from sample selection bias because we only include in our analysis the plots that were not already terraced in 1965 or before. The reason for this limitation is that in our data set we have only information on a number of village-level variables for the period after 1965. It is plausible, however, that any sample selection bias will be small, given that only 7.9 per cent of the plots had terraces by 1965.

The fourth issue is the precise specification of the different models that we estimate. We estimate three village-level models. The first is a logit model for the probability that a plot has been terraced in a village in a given year, not having been terraced before. The second is a linear regression model (OLS) for the number of plots that has been terraced for the first time in a village in a given year. The number of plots is expressed as a percentage of the total number of surveyed plots in the village. The third is a logit model for the probability that a peak in terracing has occurred in a village in a given year. The first model analyses the factors that explain the presence of agricultural intensification, the second model analyses the factors which explain the intensity of agricultural intensification, while the third model looks at the factors which explain the occurrence of a period of rapid intensification. Understanding the factors underlying these peak periods is important because more than half of the intensification took place in these peak periods (51 per cent). A peak period is defined as having more than four per cent of the village plots terraced for the first time in any given year (the average per year is 1.5 per cent for the regression period). This rule identifies the peak years discussed above, as well as two other years in which a peak occurred in one village only.[7] Table 2.3 reports the specifications for these three models.

Density (of the sub-location in which the village is located) and travel time (minutes to Nairobi by public transport) are included as village characteristics. Market-related variables are the national producer price of green coffee and white maize in KShs/Mt (in 1965 prices) and GDP per capita (in constant market prices) in Kenya. In our regressions we will include the ratio of the coffee and maize prices as a proxy for the relative attractiveness of the cash crop (coffee). The results would not change if both the price of coffee and maize were included, as only the coffee price turned out to be (positively) significant in each of the specifications. The GDP per capita variable is included as agricultural intensification may follow from increases in the demand for agricultural outputs as well as the availability of off-farm opportunities. Climate-related variables are included through variables indicating that a drought year has happened within the last three years in either Machakos or Kitui. Because there are two rainy seasons, and farmers may be able to survive one bad season, we have defined a drought year as a year in which there is a rainy season with a severe drought and a *preceding* rainy season with also a severe drought.[8] A severe drought is defined as a rainy season with a drought

---

7   These peaks occurred in 1975 (Mwanyani) and 1983 (Ngumo).

8   The two rainy seasons are the short rains (October–December) and long rains (March–May).

**Table 2.3      Model specifications**

| Category | Model 1 | Model 2 | Model 3 | Unit |
|---|---|---|---|---|
| Dependent variable | Any of the plots are terraced (dummy) | Number of plots terraced if any terracing takes place(%) | Peak year terracing (dummy) | |
| Explanatory variables | | | | |
| Village related | Density | Density | Density | persons/km |
| | Travel time to market | Travel time to market | Travel time to market | minutes |
| Market related | Price of coffee | Price of coffee | Price of coffee | Kshs/Mt |
| | Price of maize | Price of maize | Price of maize | Kshs/Mt |
| | GDP per capita | GDP per capita | GDP per capita | Current Kshs |
| Climate related | Drought year in Machakos in past 3 years | Drought year in Machakos in past 3 years | Drought year in Machakos in past 3 years | DI<=-0.8* |
| | Drought in Kitui in past 3 years | Drought in Kitui in past 3 years | Drought in Kitui in past 3 years | DI<=-0.8* |
| Unit of analysis | Village / year | Village / year | Village / year | |
| Estimation technique | Logit | OLS | Logit | |

*DI = drought index (see Tiffen et al. 1994, chapter 3). A drought year is defined as a year in which there is a rainy season with a severe drought (DI<=-0.8) and a *preceding* rainy season with also a severe drought.

index less than or equal to –0.8 (DI<=–0.8; see Tiffen et al. 1994, Chapter 3). We have also investigated the possibility of including interventions in the area in the analysis through a dummy variable indicating the presence in the past three years of a project that focuses on soil and water conservation and terracing. Unfortunately because the intervention dummy shows very little variation across time it was not possible to identify the intervention effect. All variables are time-variant except for the distance variable, which was only measured for 1998.

*Results*

The following tables give the result of the analysis. Table 2.4 presents the result of a logit analysis of the adoption of terracing in any given year. The regression in column (1) shows that the probability that any of the plots are terraced in any given year is significantly correlated with the population density, GDP per capita,

**Table 2.4**   **Logit regression analysis of whether any plot in a village is terraced in any given year (t-values in parentheses)**

| Variable | (1) | (2) | (3) | (4) |
|---|---|---|---|---|
| Density | 0.002* | 0.004 | 0.002* | 0.003 |
| | (2.11) | (1.18) | (2.04) | (0.86) |
| GDP per capita | 0.001* | 0.001 | 0.001* | 0.001 |
| | (1.81) | (1.12) | (1.74) | (1.19) |
| Travel time to market | -0.007* | | -0.007* | |
| | (2.63) | | (2.63) | |
| Price coffee/maize | 0.043* | 0.048* | 0.044* | 0.048* |
| | (2.09) | (2.17) | (1.94) | (1.94) |
| Drought in Machakos | 0.171 | 0.147 | 0.174 | 0.160 |
| | (0.39) | (0.32) | (0.38) | (0.33) |
| Drought in Kitui | 0.767* | 0.831* | 0.757* | 0.823* |
| | (1.90) | (1.89) | (1.78) | (1.77) |
| Constant | -3.001* | -3.479* | -3.097* | -3.682* |
| | (2.03) | (2.05) | (1.94) | (1.98) |
| Village effects (p-value) | No | Yes (0.20) | No | Yes (0.20) |
| Year effects (p-value) | No | No | Yes (0.26) | Yes (0.25) |
| Number of observations | 240 | 240 | 240 | 240 |
| Goodness of fit (p-value)[1] | 0.00 | 0.00 | 0.00 | 0.00 |

* Significant at 10% or lower level.

[1] Indicates the significance level of the chi-square test that none of the slope parameters are significantly different from zero.

the distance (travel time) to Nairobi and the relative price of coffee. There is also evidence that droughts in the preceding three years did motivate farmers to invest in agricultural intensification, but this effect was only found for Kitui. One explanation for this finding is that in periods of drought farmers in Machakos have better outside options in terms of off-farm employment than farmers situated in Kitui, and they are therefore less motivated to improve their farms in periods of drought.

Column (2) presents the results if village dummies are included to control for any unobserved village effects. With village dummies the cross-section patterns are subsumed and the focus is on changes within a village over time. Because our measure of distance in this model is time-invariant, we are unable to include both the village dummies and the distance variable in column (2). The coefficients do not change much from column (1) to column (2), although the significance level of the estimated coefficients suffers. We tested specification (2) against specification (1) by testing the significance of the village dummies if we also include the distance variable.[9] This test shows that specification (1) is not rejected against specification (2) (p-value 0.20). In column (3) we have included a random year variable to control for unobserved year effects and because a Hausman test did not reject random effects in favour of fixed effects.[10] Once again the coefficients are unaffected, and also the year effects are not significant (p-value 0.26). In column (4) both village and year effects are included but both are insignificant. Hence, we can accept the specification in column (1).

We tested whether lagged values of the relative coffee price were significant, but we did not find evidence for lagged price responses.[11] We also tested whether the 1970s were an exceptional period and therefore an explanation for the observed significant effect of the relative coffee price. If we interact the price variable with a dummy variable for the 1970s, we do not find that adoption reacted significantly different to relative coffee prices in the 1970s compared to the other periods. In other words, we do not detect a structural break in the model.

Table 2.5 presents the results of a linear regression (OLS) analysis explaining the intensity of terracing in any given village and year. The intensity of terracing is measured as the percentage of plots that has been terraced. None of the population density, the relative price of coffee, distance and drought variables appear as significant. In column (2) we have also included a sample selectivity term to control for the fact that the regression only includes positive observations.[12] This does not change the results. We have also tried to include village and year effects but they were insignificant and did not change the results. Hence, the results show that although the population density, the relative price of coffee, distance and drought variables can explain the probability that any terracing takes place in the village (Table 2.4), they are unable to explain the intensity of terracing.

The fact that we do not find any significant effect of the village-level variables on the intensity of terracing may be due to a non-linear relationship between these variables. If there are any fixed costs to terracing, then farmers may prefer to terrace in 'bunches' or peak periods. For instance, if terracing involves the mobilization of exchange labour groups, then it is easier to concentrate terracing activity in those years in which these exchange groups are most active. Hence, although we may

---

9     More precisely, we included six of the eight village dummies because one village dummy was dropped because of the adding up constraint and another dummy because the (village-specific) distance variable is included.

10    See Greene (2000), Chapter 14.

11    We included three lags for coffee and maize prices, but they turned out to be insignificant in a joint $\chi^2$-test.

12    Heckman and MaCurdy (1986, table 1) show that if the probability that terracing occurs is specified as a logit model, that the sample selectivity term is given by $-[\ln F(X\beta)-X\beta F(-X\beta)]/F(X\beta)$, where F is the cumulative logistic distribution, X the vector of explanatory variables, and $\beta$ the corresponding vector of coefficients. The sample selectivity term is calculated by using the estimated vector of coefficients from the logit model (Table 2.4, column (1)).

**Table 2.5    OLS regression analysis of percentage of village plots terraced if any terracing takes place in a village in any given year (t-values in parentheses)**

| Variable | (1) | (2) |
|---|---|---|
| Density | -0.001 | -0.004 |
| | (0.49) | (0.57) |
| GDP per capita | -0.0006 | -0.002 |
| | (1.00) | (0.70) |
| Travel time to market | -0.003 | 0.011 |
| | (0.07) | (0.45) |
| Price coffee/maize | 0.033 | -0.037 |
| | (0.95) | (0.24) |
| Drought in Machakos | -0.759 | 1.047 |
| | (1.07) | (1.11) |
| Drought in Kitui | 0.606 | -0.713 |
| | (0.89) | (0.24) |
| Sample selectivity term | | -2.443 |
| | | (0.47) |
| Constant | 5.24* | 13.87 |
| | (1.98) | (0.74) |
| | | |
| Number of observations | 107 | 107 |
| Goodness of fit (p-value)[1] | 0.43 | 0.52 |

[1] Indicates the significance level of the chi-square test that none of the slope parameters are significantly different from zero.

not find any effect of the village-level variables on terracing intensity in general, we may find an effect if we concentrate on those peak periods.

Table 2.6 presents the results of a logit analysis of the occurrence of peaks in terracing in any given village and year. A peak in terracing is observed if at least four per cent of the plots in the village are terraced in a given year (the average percentage is 1.5 per cent). We have also tried to include village and year effects but they were insignificant and did not change the results. Travel time to the market and drought in Kitui are the only significant explanatory factors for the occurrence of peaks. The effect of the coffee price is positive but not significant. However, if we define a drought year as any calendar year in which there is less than 800 mm of rainfall, then the coffee price also significantly explains the historical peaks in terrace construction. Villages that were lying further from the market were less likely to experience any peaks in terracing. The latter result may be explained by the presence of transaction costs – villages that were further from the market could not benefit as much from the boom in coffee prices because of higher transportation and/or information costs. Drought in Machakos does not appear as a significant

**Table 2.6**     **Logit regression analysis of whether there is a peak in the number of plots terraced in a village in any given year (t-values in parentheses)**

| Variable | |
|---|---|
| Density | 0.001 |
| | (0.51) |
| GDP per capita | -0.00003 |
| | (0.05) |
| Travel time to market | -0.010* |
| | (2.44) |
| Price coffee/maize | 0.037 |
| | (1.30) |
| Drought in Machakos | -1.152 |
| | (1.41) |
| Drought in Kitui | 1.019* |
| | (1.64) |
| Constant | -3.095 |
| | (1.44) |
| Number of observations | 240 |
| Goodness of fit (p-vale)[1] | 0.04 |

[1] Indicates the significance level of the chi-square test that none of the slope parameters are significantly different from zero.

variable, presumably because farmers in Machakos have better outside options in terms of off-farm employment than farmers in Kitui.

*Do omitted household and plot characteristics bias our results?*

One may question the above results because we have not controlled for household and plot characteristics. The reason for this omission is that we do not have retrospective data on household and plot characteristics other than terrace construction. We have, however, included village dummies to control for time-invariant village-level differences in household and plot characteristics, and we have seen that the results remain unaffected. We also saw that the results are robust to the inclusion of year effects. Here we will test for the significance of household and/or plot heterogeneity by estimating a model of adoption at the *plot-level*, instead of at the village-level. If the results of this analysis are similar to those found earlier, we may conclude that omitted household and plot characteristics do not drive the earlier results. It should be pointed out that the plot-level analysis itself also suffers from a number of weaknesses. It does not include the effects of village interactions such as copying effects, technological spillover and endogenous village prices, because there is no simple way to model them. Also time-varying household and plot characteristics are omitted because of lack of data. Hence, we use the results of the plot-level analysis to see whether the earlier results are robust to the inclusion of time-invariant household and plot characteristics, but not necessarily as a superior model of terrace adoption.

The specification of the plot-level model follows the adoption model developed in Pomp and Burger (1995). The logit model gives the probability of adoption in a given year for a given plot. Each year any plot that has not been terraced before has a probability of being terraced. If terracing occurs, then the plot drops out of the model. If terracing does not occur, then the plot gets another chance in the next year. The logit model includes the same village-level explanatory variables as in the earlier analyses, as well as a number of other variables to control for household and slope heterogeneity: dummies for slope and soil type, a variable measuring the distance from plot to home in metres, and household fixed effects.[13] Note that with our data we can only estimate a model for adoption at the plot level, and not for plot-level adoption intensity or the presence of peaks in adoption activity. Also the travel time to the market cannot be included because of the presence of household fixed effects. Table 2.7 presents the results of the analysis.

Initially we included several slope and soil-type dummies, but none of the soil-type dummies turned out to be significant, and for the slope dummies we did not find a significantly different effect between lower, medium and upper slope. Hence, in the regressions we have combined these three slope types. The most important result from the Table 2.7 is that each of the significant village-level variables in Table 2.4 is also significant in Table 2.7, except the drought variable (column (1)). The plot-slope variable is also significant and positive, as expected. This result would suggest that the omission of household and plot characteristics does not bias the results in any serious way as we can still conclude that density, GDP per capita and coffee price are significant factors in the adoption of terracing. The size of the coefficients are not directly comparable however because of the aggregation problem. If year effects are included (column (2)) then the coefficient of the density variable becomes much smaller and strongly insignificant.[14] These year effects also turn out to be strongly significant in this case. In the village-level model (Table 2.4) we did not find that the density variable became insignificant if year effects were included. This suggests that the village-level density variable is a good proxy for population pressure in the village-level model, but not in the plot-level model. Population pressure felt at the plot level depends as much on village-level population pressure as on household characteristics (land ownership and household size), and therefore may vary strongly between households and plots.

---

13    A Hausman test clearly rejected random effects in favour of fixed effects.

14    We have included random year effects because the Hausman test does not reject random year effects against fixed year effects.

**Table 2.7**    **Logit regression analysis of whether any plot in a village is terraced in any given year. Controlling for household and plot heterogeneity (t-values in parentheses)**

| Variable | (1) | (2) |
|---|---|---|
| Density | 0.011* | 0.001 |
|  | (5.81) | (0.05) |
| GDP per capita | 0.001* | 0.003* |
|  | (4.29) | (7.14) |
| Price coffee/maize | 0.028* | 0.028 |
|  | (2.50) | (1.04) |
| Drought in Machakos | -0.216 | -0.599* |
|  | (0.87) | (1.83) |
| Drought in Kitui | 0.936* | 0.116 |
|  | (3.39) | (0.34) |
| Distance plot to home | -0.0001 | -0.00004 |
|  | (1.37) | (0.69) |
| Lower/medium/upper slope | 1.191* | 1.700* |
|  | (2.44) | (3.31) |
|  |  |  |
| Household effects | Yes | Yes |
| Year effects | No | Yes |
| (p-value) |  | (0.00) |
| Number of observations | 5752 | 5752 |
| Goodness of fit (p-value)1 | 0.00 | 0.00 |

[1] Indicates the significance level of the chi-square test that none of the slope parameters are significantly different from zero.

*Population pressure or other factors?*

How much of the Machakos miracle can be attributed to increases in population pressure over the sample period? The above analysis shows that besides population pressure, coffee prices, droughts and travel distance to the capital city also play a role in explaining the Machakos miracle. In fact, the above analysis allows us to analyse the relative importance of population pressure against these other factors.

We have used the village-level model to simulate what would have happened to the *cumulative rate of adoption* over the period 1966–95 if population density had remained at the level observed in 1966 in each of the villages.[15] As a contrast, we also have simulated what would have happened if there had been no droughts, if there had been no coffee boom (the relative coffee price would have stayed at levels observed for 1966) or if the travel times to the villages had been twice as

---

15    For the village-level model the cumulative rate of adoption is calculated as the weighted average of the rate of village-level adoption with weights equal to the share of plots found in each of the villages.

long as currently observed.[16] Figure 2.6 shows the cumulative patterns of adoption under each of these scenarios, using the village-level results from column (1) in Table 2.4.[17]

The simulations for population density, coffee price boom and especially the reduction in travel time to Nairobi show the largest effect on agricultural intensification in Machakos and Kitui. If travel time from Machakos and Kitui had been double, then the models predict that the cumulative rate of adoption would be at most 32 per cent in 1995, as opposed to the observed 55 per cent. The coffee boom appears to have been the second strongest factor in the model. If coffee prices had stayed at their 1966 (real) price level, then the cumulative adoption rate would have been 39 per cent. For density we find that if the population density

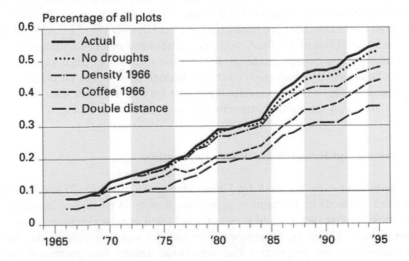

**Figure 2.6     Model simulations of cumulative adoption rates for all villages using village-level model, 1966–95**

16    A halving of the travel times between 1966 and 1995 appears to be a conservative estimate of the improvement in transportation infrastructure over this period, and the effect of this may have been more dramatic for Kitui than for Machakos due to the relatively recent tarring of the Machakos–Kitui road.

17    The terracing intensity in a given village and year is predicted by the probability that any terracing took place (from Table 2.4, column (1)) times the average reported terracing intensity for villages and years that any terracing was reported (3.4 per cent). The results for the travel time and coffee price simulations barely change if we use the model for peak years instead. In that case the terracing intensity in a given village and year is predicted by the weighted average of the average reported terracing intensities in peak (6.7 per cent) and off-peak years (2.2 per cent) for villages and years that any terracing was reported, with weights equal to the predicted probabilities that a peak or off-peak period occurred (from Table 2.6).

had remained at the level observed in 1966, then the cumulative adoption rate would have been 42 per cent. The effect of droughts is found to be of relatively minor importance in the model simulation, explaining 8 per cent of the cumulative terracing rate in 1995.[18]

Summarizing, the model simulation in Figure 2.6 suggests that the improvement in infrastructure was a driving force behind the Machakos/Kitui miracle. The rise in population density and the coffee boom played also a role but their effect appears to have been weaker. Of course here we have analysed the direct impact of population density and not the indirect impact. For instance, development of transportation infrastructure may itself be a function of population density and as such we may have *under*estimated the role of population pressure for agricultural development. This might indeed be true, but the opposite might also hold – population pressure may itself be affected by other factors, such as transportation infrastructure, and we might even have *over*estimated the role of population pressure. In addition, a positive correlation between population and road density should not be taken for granted either (Pender 1999). A much more elaborated model would be necessary to study all these linkages, to assess the ultimate role of population pressure, and, to our knowledge, no such study has been undertaken as of yet. It is clear however that other factors besides population pressure may have had at least as strong a *direct* impact on agricultural intensification as population pressure.

### Discussion and Research and Policy Implications

Even though the earlier work by Tiffen, Mortimore and Gichuki and their team focused on the district's trend in population density in particular, attention was also paid to other factors conducive to the adoption of natural resource enhancing technologies. However, no retrospective data set was available at the time to rigorously test the proposition that population density was paramount among the variables explaining the adoption of these techniques, as this needed a study at the combined plot, household and village level. Of course, even with peaks in adoption made possible by high cash crop prices, ultimately the correlation between population density and terracing would have ended up as being high. The present research offers the first opportunity to test the hypotheses against the evidence.

The result of this test points at the direct impact of low transaction costs, as operationalized by the time it takes to arrive at the major market, next to population density itself. Both the regular and easy flow of people (farmers, researchers and extension officers alike) and the resultant information flow, and the reduced costs of the transport of inputs and output may have stimulated the construction of terraces. The fact that the windfall profits of the coffee boom seem to have been used extensively for this purpose points at the direct link between cash crop needs

---

18    Without any droughts the cumulative adoption rate would be 47 per cent.

(in terms of moisture and nutrient management) and profitability (resulting in the necessary financial resources for the actual implementation of terraces). This link is strengthened through reduced transaction costs. Continued work is now done to further test this idea through an analysis of crop choice and profitability of the terracing technique (Oostendorp and Zaal in this volume; Mwakubo et al. in this volume). Another impact of transaction costs reduction may be the facilitation of seasonal work in the major cities of Kenya. Much of the explanatory force of this argument focuses on the level of the household, as adoption depends partly on labour availability and wealth. For this reason, another analysis is now being done to determine the factors explaining the variance at the household level. In addition the plot level is considered, as much of the explanation is at that level as well.

Interestingly, droughts do not appear the prominent forces of change as they were assumed to be beforehand. The finding that, in Machakos, this variable is even less significant as it is in Kitui further corroborates this. The hypothesis is that the relatively low costs of seasonal migration and the higher level of economic activity generally facilitate alternative ways to secure a livelihood apart from further investments in agriculture. This in itself could indirectly lead to higher levels of adoption of inputs and terracing in Machakos, as market opportunities could be seized independently of drought conditions.

In terms of policy, the results point at the need to consider transaction costs and in particular infrastructure development, in relation to policies on direct transmission of benefits from cash cropping and any credit facilities that would further enhance the proper use of these profits for investments in the quality of land by small farmers. The construction boom noticed by Bevan et al. (1992) partly found its expression in investments in land quality through terrace construction, but the timing of these investments would have, if deferred to a later date, facilitated the choice of a period with lower labour costs. Consequently, more terraces would have been constructed. On the other hand, this boom certainly caused a take-off phase, accelerating the process of agricultural intensification in Machakos and Kitui.

Chapter 3

# Farm- and Household-Level Drivers of Agricultural Innovation in Machakos and Kitui Districts, Kenya

Remco Oostendorp and Fred Zaal

## Introduction

In the previous chapter we studied the drivers that increase both productivity and the maintenance or increase of levels of fertility and water availability at the village and regional (district) level. We did this because the initial discussion on the relationship between population growth and rural agricultural production focused on this level (Boserup 1965; Tiffen et al. 1994). However, other levels of explanation should be considered as well, as the adoption of soil and water conservation techniques not only depends on village and regional factors, but also on household- and plot-level factors. In this chapter we aim to do that: to derive insights from the process of conservation adoption at multiple levels. A more balanced picture could provide the basis for more effective and efficient policies that focus on productivity growth and sustainable agriculture.

## Conservation Farming and the Level of Explanation

Soil and water conservation, the main component of conservation farming, takes place at the plot level but it is well-known that the actual adoption decision depends not only on plot-level characteristics, such as soil type, soil fertility, soil moisture levels and plot slope, but also on factors operating at the household, village or higher level. The literature on non-separable farm household models suggests that household-level variables such as the number of family members and the availability of cash or access to credit may have a significant impact on soil and water conservation in the presence of market imperfections (Pender and Kerr 1998; Shiferaw and Holden 2000). Studies emphasizing the role of transaction costs and population density often analyse village-level variables such as distance to the main market and the number of people per square kilometre in the village (Boserup 1965; Tiffen et al. 1994; Jacoby 2000; Fafchamps and Shilpi 2003; Ersado et al. 2004; Zaal and Oostendorp this volume; Mwakubo et al. this volume).[1] Studies emphasizing the role of policy on soil and water conservation consider provincial-, district- or even national-level factors, such as prices and tariffs (Coxhead and Jayasuriya 1995; Pagiola 1996).

In the following, we aim to study these factors at the various levels. When we speak of 'village-level characteristics', we mean characteristics that are defined at the village level (for example the percentage of plots terraced in a village). This does not imply that a village-level characteristic is (fully) determined by village-level action – it may depend on action at the collective (district, village) as well as individual (household) level, and it also depends on plot characteristics (see the discussion on the interdependence of the different levels below).

There are large variations at the village level in the adoption of soil and water conservation techniques. Using the survey for Machakos and Kitui, we calculate adoption rates at the plot level of various techniques in the villages studied, and distinguish among the following conservation techniques: use of manure, use of artificial fertilizer, use of green manure or compost, terracing, use of trash lines, use of grass strips, and use of grassed terrace borders. Table 3.1 shows that, on average, terracing is the most commonly used conservation technique, with 66 per cent of the plots being terraced. The next most frequently used technique is manuring, on 47 per cent of the plots. The other conservation techniques are only used on a small minority of the plots.

These average figures hide significant variation across villages however. The terrace adoption rate varies from only 30 per cent in the village Utwiini to 79 per cent in the village Ngalalia. The adoption rate of manuring varies from 19 per cent

**Table 3.1**    **Adoption rates of soil and water conservation techniques across villages, 1999**

| Village | Manure | Artificial Fertiliser | Green manure and compost | Terraces | Trash lines | Grass strips | Grassed terrace border |
|---|---|---|---|---|---|---|---|
| Kisaki | 0.62 | 0.23 | 0.04 | 0.77 | 0.01 | 0.21 | 0.39 |
| Musoka | 0.59 | 0.22 | 0.08 | 0.69 | 0.08 | 0.16 | 0.10 |
| Ngalalia | 0.60 | 0.51 | 0.19 | 0.79 | 0.09 | 0.16 | 0.09 |
| Ngumo | 0.34 | 0.13 | 0.10 | 0.69 | 0.00 | 0.05 | 0.05 |
| Mwanyani | 0.37 | 0.05 | 0.03 | 0.58 | 0.07 | 0.27 | 0.00 |
| Kitungati | 0.19 | 0.00 | 0.00 | 0.43 | 0.09 | 0.02 | 0.08 |
| Utwiini | 0.30 | 0.03 | 0.06 | 0.30 | 0.45 | 0.00 | 0.09 |
| Kyondoni | 0.58 | 0.00 | 0.04 | 0.70 | 0.06 | 0.17 | 0.00 |
| Total | 0.47 | 0.19 | 0.08 | 0.66 | 0.08 | 0.14 | 0.11 |

1   In microeconometric studies the distance variable typically reflects not only the isolation of the village (a village-level variable) but also the isolation of the household within the village (a household-level variable).

in the village Kitungati to 62 per cent in the village Kisaki. Similar large village-level variations can be found for the other conservation techniques.

These large village-level variations suggest that village-level factors, such as local population pressures, distances to local marketplaces and the presence of local institutions (such as extension services, NGOs, labour-sharing institutions) may be able to explain largely why some villages show higher levels of soil conservation investments. However, some of this variation may already occur at the district level, and adoption rates probably also vary significantly within a village across households and even across plots within a household.

For this reason many studies on the adoption of soil and water conservation techniques consider multiple levels of explanation by including plot-level, household-level and higher-level variables in the analysis (Pender and Kerr 1998; Adégbidi et al. 2004). Typically a number of proxy variables are constructed that are thought to capture the main determinants of adoption at each level. It would be interesting and useful, however, to ask a more general question first: *how much of the variance in adoption of soil and water conservation techniques can in principle be found at each level of explanation?* In other words, is most of the variation in adoption behaviour found across different villages, or across different households within a given village, or across different plots within a given household? In case most of the variation is found across villages, this could suggest that village-level factors are most important for understanding adoption behaviour. At the other extreme, if most of the variation is found across plots for a given household, then plot-level characteristics may play a dominant role in the adoption process.

Obviously knowledge of the level at which most of the variation in adoption occurs will be helpful for researchers and policy-makers to target further analysis and interventions at the appropriate level. If most of the variation in adoption is found at the plot level, then an improvement of the road between the village centre and the main (outside) market will probably have limited impact on adoption. However, improvements in soil and water conservation technologies allowing marginal plots to become more productive can be expected to have a major impact.

It is important, however, to realize that in reality the different levels are not independent of each other. For instance, if villagers improve the village tracks from the village centre to the different plots, then not only the travel distance from the plot to the village centre is reduced (a plot-level characteristic), but also the average distance of all plots in the village to the village centre (a village-level characteristic). Similarly, if a new road is built between the village centre and the outside main market, then travel time between the village and the main market is reduced (a village-level characteristic) but also the travel time between each individual plot and the main market (a plot-level characteristic). It is therefore conceivable that, say, most of the variation in adoption behaviour is found at the village level, but that plot-level (or household-level) interventions are most effective to change a village-level characteristic. Although it is plausible that in most cases the most effective intervention is done at the same level, in practice this will depend on the interrelationships between the different levels and the relative cost at which different interventions at different levels can be introduced.

Keeping the above caveat in mind, we proceed with an analysis of variance for different types of soil and water conservation techniques in Machakos and Kitui. We apply a sequential analysis of variance with four levels of analysis, namely the district, village, household and plot levels. First we calculate how much of the total variance can be explained at the district level. Second, we calculate how much of the variance can be explained additionally at the village level. Third, the additional variance explained at the household level is calculated. Fourth, and finally, the remaining variance at the plot level is computed. As this is a standard ANOVA analysis, there is no residue that remains unexplained.

The first column in Table 3.2 reports the total variance in soil conservation adoption by conservation type across all districts, villages, households, and plots. In the next column we report the percentage explained of this total variance (by type) by district-level variation. In principle if all adopters could be found in one of the districts, this percentage would be 100. On the other hand, if the adopters are spread equally across districts, then the percentage would be zero. The third column shows how much of the variance can be explained by village-level variation, in addition to what can already be explained at the village-level. The fourth column shows how much of the variance can be explained by household-level variation, in addition to what has already been explained by the district and village-level. The last column shows the remaining part of the variance, at the plot level, which cannot be explained at the district, village or household level.

The table shows clearly that most of the variation in soil conservation adoption rates can be explained at primarily the household and subsequently the lower (plot) level. The total variance explained by district- and village-level characteristics is less than a quarter. It is highest for the use of artificial fertilizer, trash lines and grass terrace borders. Interestingly enough, the variation in adoption of terraces consists for less than 10 per cent at the district and village-level, and for more than 50 per cent at the plot-level. This makes terracing exceptionally plot-dependent

**Table 3.2    Sequential Analysis of Variance of soil and water conservation techniques**

| Type | Total variance | District (%) | Village (%) | Household (%) | Plot (%) |
|---|---|---|---|---|---|
| Manure | 0.25 | 3.03 | 6.43 | 51.96 | 38.59 |
| Artificial fertiliser | 0.15 | 12.77 | 9.69 | 50.79 | 26.76 |
| Green manure and compost | 0.07 | 2.22 | 3.03 | 60.71 | 34.03 |
| Terraces | 0.23 | 5.33 | 4.08 | 38.76 | 51.82 |
| Trash lines | 0.08 | 2.38 | 11.95 | 52.58 | 33.09 |
| Grass strips | 0.12 | 0.05 | 5.79 | 63.85 | 30.31 |
| Grassed terrace border | 0.10 | 3.75 | 12.32 | 58.65 | 25.28 |

– for the other soil conservation techniques the total variance explained at the plot level is one third or less.

Can we therefore conclude that the Machakos miracle is more the result of a specific combination of household and/or plot-specific characteristics rather than the more general village- or district-level characteristics? Possibly. Still, there are two issues to consider. First, the villages in the survey may not be representative for all villages in Machakos and Kitui because only eight villages are included and villages were selected according to population density and distance to Nairobi in travel time (see Chapter 2). Therefore the above ANOVA results are specific for our sample and may not be generalizable to the whole population of villages in Machakos and Kitui. However the lack of district- and village-level variation cannot be attributed to the fact that the surveyed villages are simply quite similar. Table 3.1 already showed a large variation in adoption across the villages and the history and situation of the two districts, Machakos and Kitui, are very different also, with Kitui having a much lower population density and being much farther from the important Nairobi market than Machakos.

Second, one may argue that the above figures reflect the current situation, and that it is only now, after decades of agricultural intensification, that household- and plot-level characteristics become increasingly important. Presumably, according to this line of thinking, at the onset of agricultural intensification, higher-level factors such as population pressure and closeness to market centres played a crucial role. Later, with increasing intensification, other, more household- and plot-specific factors (such as labour shortages and lack of need for terracing) become more important.

In light of this it is interesting to repeat the sequential analysis of variance for earlier time periods. In case it is still found that most of the variation is at the household and plot level, this suggests that the dominant role of household- and plot-level factors is not period-specific (e.g. 1970s versus 1990s) nor valid only at a specific phase of village-level transition (e.g. early stage of transition versus late stage of transition).

To repeat the analysis of variance for earlier periods, we have used the data on the year when the soil conservation technique was first introduced on the plot. This information is available for four types of soil and water conservation techniques, namely terracing, use of trash lines, use of grass strips and use of grassed terrace borders. We distinguish the following years: 1960, 1970, 1980, 1990 and 1999 (the survey year). In this manner we obtain five 'snapshots' of the soil conservation landscape over the past four decades. In Table 3.3 we report the adoption rates for the different techniques across the different years. Terracing was the first soil conservation technique adopted, followed by grass strips and the use of grassed terrace borders.

In Table 3.4 we have decomposed the total variance in adoption rates of terracing across the years. We limit this analysis to terracing because the previous table shows that the adoption rates for the other techniques were very low in the earlier years. Once again we find that household and plot characteristics remain the

**Table 3.3      Cumulative adoption rates of soil and water conservation techniques, 1960–99**

| Year | Type | | | |
|------|---------|-------------|-------------|----------------------|
|      | Terrace | Trash Lines | Grass strip | Grassed terrace border |
| 1960 | 0.06 | 0.00 | 0.00 | 0.00 |
| 1970 | 0.13 | 0.00 | 0.02 | 0.02 |
| 1980 | 0.29 | 0.02 | 0.05 | 0.03 |
| 1990 | 0.48 | 0.04 | 0.07 | 0.06 |
| 1999 | 0.66 | 0.08 | 0.14 | 0.11 |

**Table 3.4      Sequential Analysis of Variance of terracing, 1960–99**

| Year | Total variance | District (%) | Village (%) | Household (%) | Plot (%) |
|------|---------|--------------|-------------|---------------|----------|
| 1960 | 0.05 | 3.46 | 1.34 | 71.16 | 24.04 |
| 1970 | 0.11 | 6.08 | 3.45 | 57.86 | 32.60 |
| 1980 | 0.21 | 6.24 | 2.30 | 54.47 | 36.99 |
| 1990 | 0.25 | 8.21 | 2.14 | 48.22 | 41.43 |
| 1999 | 0.23 | 5.33 | 4.08 | 38.76 | 51.82 |

dominant factors accounting for most of the variance in soil conservation adoption rates, even for the initial years of adoption. Still there is an interesting shift over time – household characteristics are dominant at the early stage of adoption, while plot characteristics are dominant at later stages of adoption. At the early stage of transition most terracing takes place in a limited number of households, while at later stages, terracing takes place on a limited number of plots in most households.

The earlier question, whether household and plot characteristics are more important and have been across time periods, can be answered in the affirmative. This is a rather sobering thought, and it puts the focus on district-level drivers in perspective.

**A Multi-level Model of Adoption of Terracing, Manuring and Fertilizing**

The previous section has shown that the variance in adoption of soil and water conservation techniques is located at multiple levels, mostly at the household and plot levels, but also at the village and district levels. This finding, together with the literature emphasizing different levels of explanation, suggests that any empirical model of soil and water conservation adoption should include plot-level, household-level and higher-level variables in the analysis.

In this section we develop a multi-level model to understand the important plot-level, household-level and higher-level factors affecting the decision to terrace and to use manure and fertilizer. The model here follows the framework of Pender and Kerr (1998). They developed a simple investment model of soil and water conservation under imperfect credit and/or labour markets. The main implication of the model is that soil and water conservation investments are unaffected by the farmers' labour and capital endowments. However with an imperfect credit and/or labour market, investments may not only depend on factors affecting the profitability of an investment, but also on farmers' labour and capital endowments. This is the well-known Fisher Separation Theorem applied to investments in soil and water conservation.

Pender and Kerr estimate the plot-, household- and village-level determinants of soil and water conservation for a number of ICRISAT villages in India.[2] Here we estimate the determinants of the probability of soil and water conservation in Machakos and Kitui, particularly terracing, manuring and fertilizing. The dependent variables are whether the plot is terraced, manured or fertilized respectively. The explanatory variables are grouped in plot-, household- and village-level variables.

The plot-level variables capture the plot characteristics that may have an impact on investment. Distance (in metres) between the plot and the house of the owner/user was included as a measure of transaction costs within the farm. Plots that are farther away from the house are expected to receive less soil and water conservation investments. Plot size is included because we expect that larger plots are more likely to receive investments if there are fixed costs of investments. The tenure status of the plot is included as plots with more secure tenure are expected to receive more investments as well. We include two dummies for the tenure status, namely whether there is a private title deed or whether the farmer is still obtaining the title deed (but the land is demarcated). The omitted category includes traditional rights based on past (intergenerational) record of use or communal rights, and a few instances of informally occupied land without traditional or formal rights, squatting and rented land. The soil type of the plot will affect the profitability of soil and water conservation investments, and therefore dummies were added for sandy and sandy/loam soils and for loamy/silt soils (omitted category: clay and other soils). Similarly the profitability of investment may vary with the slope of the plot, and we included dummies for lower slope, medium slope and upper slope.

Among the household variables, the other category, the size of the household as well as the share of household members in the age classes 18–39 years, 40–59 years, and 60 and older (with omitted category the share of household members below 18 years) are included. Larger households and households with a larger share of young and old members have a smaller supply of (effective) labour and

_____

2 The village-level variables are captured by village dummies (in the pooled regressions for all villages).

are expected to invest less, for instance because of labour market imperfections. The highest education level of any of the members of the household is included as the knowledge and skill of the household may affect the awareness and profitability of soil and water conservation investment. Two dummies are included for the education level, namely for secondary and for college education completed (omitted category is no secondary schooling completed). Also, the proportion of household members between 18 and 60 years engaged in off-farm or self-employed rural or urban activities (as a main occupation) was added. Off-farm and self-employment may relieve the credit constraint for investment in case credit markets are imperfect, and therefore can have a positive impact on soil and water conservation. We also include two measures of wealth, namely the total area of land owned or used and total livestock (cattle and smallstock in tropical livestock units. One cow equals 0.7 TLU, one goat or sheep equals 0.1 TLU) as richer farmers will be more likely to invest if credit markets are imperfect. Of course, the wealth of households may also have increased because of investments in the past, and therefore the coefficients on these wealth variables should be interpreted with caution.[3] In the survey, the livestock information was sometimes missing but it is not always clear whether this reflects a total absence of livestock or simply missing observations. In these cases we set the total livestock variable equal to zero, but we include a dummy variable that equals one if no livestock information is available.

With respect to the village-level variables, we included village dummies to capture any of the village or higher-level factors affecting soil and water conservation investments. Of 484 plots (and 193 households), a total of 369 were complete on all variables, and these were used in the model. Table 3.5 reports the descriptive statistics. In the sample, 72 per cent of the plots are terraced, while respectively 53 and 21 per cent of the plots receive manuring or fertilizer. Plots are on average 1.94 acres and almost 1 km away from the home. About a third of the plots have private title deed, and almost half are awaiting this deed. Households have on average almost seven members, and are relatively young – 37 per cent of the household members are between 18 and 39 years and 42 per cent below 18 years. More than half of the households have no member with secondary education completed or higher, but 20 per cent of the households have a member with college

---

3   For example, the level of wealth may have increased because of investments in terracing in the past, as we take all terraced plots into account, and terracing started some sixty years previously (we thank Kees Burger for making this point). This causes a circular argument that can be avoided when we take recently terraced plots only. However, a selection bias may then be introduced based on the factors that made some plots be terraced earlier than others. As older terraces were more likely built by selected and probably more wealthy and better-informed households (we refer to Table 3.4 which showed the initial high explained variance at household level of terracing) we assume our argument remains valid. It does however point again at the importance of the time dimension in the analysis of adoption.

**Table 3.5     Descriptive statistics**

| Variables | Mean | St. Dev. | Min | Max |
|---|---|---|---|---|
| **Dependent variables** | | | | |
| Terracing (dummy) | 0.72 | 0.45 | 0.00 | 1.00 |
| Manuring (dummy) | 0.53 | 0.50 | 0.00 | 1.00 |
| Fertilizing (dummy) | 0.21 | 0.41 | 0.00 | 1.00 |
| **Plot characteristics** | | | | |
| Distance plot to home (m) | 919.65 | 3024.49 | 0.00 | 40000.00 |
| Plot size (acres) | 1.94 | 3.25 | 0.05 | 40.00 |
| Tenure | | | | |
| Private title deed (dummy) | 0.34 | 0.48 | 0.00 | 1.00 |
| Still obtaining title deed/demarcated (dummy) | 0.46 | 0.50 | 0.00 | 1.00 |
| Slope | | | | |
| Lower slope | 0.46 | 0.50 | 0.00 | 1.00 |
| Medium slope | 0.26 | 0.44 | 0.00 | 1.00 |
| Upper slope | 0.12 | 0.32 | 0.00 | 1.00 |
| Soil type | | | | |
| Sandy/sandy/loam | 0.62 | 0.49 | 0.00 | 1.00 |
| Loamy/silt | 0.24 | 0.43 | 0.00 | 1.00 |
| **Household characteristics** | | | | |
| Household size | 6.95 | 2.62 | 2.00 | 16.00 |
| Age composition (proportion) | | | | |
| 18–39 years | 0.37 | 0.18 | 0.00 | 1.00 |
| 40–59 years | 0.16 | 0.15 | 0.00 | 0.67 |
| 60 or more years | 0.05 | 0.11 | 0.00 | 0.50 |
| Highest education in household (dummy) | | | | |
| Secondary school | 0.20 | 0.30 | 0.00 | 1.00 |
| College | 0.20 | 0.33 | 0.00 | 1.00 |
| Off-farm or self-employment (proportion) | 0.34 | 0.31 | 0.00 | 1.00 |
| Total land owned or used (acres) | 6.13 | 7.98 | 0.25 | 51.00 |
| Animal cattle and smallstock (TLU) | 4.35 | 5.71 | 0.00 | 37.60 |
| No information on livestock (dummy) | 0.09 | 0.28 | 0.00 | 1.00 |
| **Village characteristics** | | | | |
| Musoka (dummy) | 0.11 | 0.31 | 0.00 | 1.00 |
| Ngalalia (dummy) | 0.23 | 0.42 | 0.00 | 1.00 |
| Ngumo (dummy) | 0.12 | 0.32 | 0.00 | 1.00 |
| Mwanyani (dummy) | 0.15 | 0.35 | 0.00 | 1.00 |
| Kitungati (dummy) | 0.08 | 0.28 | 0.00 | 1.00 |
| Utwiini (dummy) | 0.07 | 0.26 | 0.00 | 1.00 |
| Kyondoni (dummy) | 0.11 | 0.31 | 0.00 | 1.00 |

*Source*: Authors' analysis based on NWO survey data.

education or higher. About a third of the household members between 18 and 60 years are involved in off-farm or self-employment as their main occupation.

The regression models are estimated with logit and the results are presented below. Among the plot-level variables, only the tenure status affects the probability of terracing. The effects are in the expected direction – plots with a private title deed or still obtaining one are 16.1 and 28.5 per cent more likely to be terraced respectively – confirming that plots with more secure status receive more investments.[4] For manuring a large number of plot-level characteristics play a role – plots that are further from the house are less likely to be manured, higher slopes receive more manure, and manuring is much more likely on sandy, sandy/loam and loamy/silt soils than on clay and other soils (the omitted category). For fertilizer only one of the plot-level characteristics was found to be significant – loamy/silt soils have a 42.8 per cent higher probability of receiving fertilizer.

A number of household characteristics affect the probability of terracing significantly. Plots in larger households are less rather than more likely to be terraced, contrary to expectations. Maybe larger households tend to be poorer and, although they have more hands to construct terraces, they lack the cash to pay for complementary inputs (such as fertilizer, seeds and seedlings). Alternatively, they are more likely to be stuck in a Malthusian poverty trap, where they focus on the cultivation of current crops rather than farm system improvements for the future. A third explanation may be found in the idea that larger households are also often older households, and that this may be an effect of the household life cycle rather than a poverty-related effect. This is a time-dimension issue to which we will turn below. In terms of age structure, households with a large proportion of members between 40 and 59 years are much less likely to construct terraces than households with relatively younger members. This suggests that there is a 'generational divide' between the different households in terms of terracing, and in the next section we will find further evidence for this. However, looking at the coefficient for the share of older household members (60 years and older), the results also suggest a non-linear relationship with age structure, but the dummy for age 60 or more years is not significant. Schooling has a large impact on terracing – plots in households with at least one member with a secondary or higher education are about 30 per cent more likely to be terraced. Although off-farm and non-farm employment and total land do indicate a significant impact on the probability of terracing, households with more livestock construct more terraces as expected.

With respect to manuring, none of the included household characteristics play a role, suggesting that manuring is a highly plot-specific activity. This, however, is in contradiction with the sequential analysis of variance, which showed that

---

4   It is interesting that the effect is larger for plots for which the title deed is still being obtained. Given that these plots, on average, have been acquired more recently than plots already under private title deeds, this suggests that acquisition has an independent positive effect on investment, apart from the tenure status. The dynamic analysis later in this chapter confirms this conjecture.

household characteristics play a relatively important role for manuring (Table 3.2). This suggests that the included household characteristics do not capture a number of important omitted (and unobserved) household characteristics, for instance prior experience with manuring and disutility of working with manure. With respect to artificial fertilizing, only a few household characteristics are found to play a role, namely education and off-farm or self-employment. Plots of households with college-educated members and households with off-farm or self-employment are 26–38 per cent more likely to be using artificial fertilizer, suggesting the importance of education and cash income. Village dummies are also included in the regression and they are found to be jointly significant at the 1 per cent significance level for terracing and fertilizer, and at the 10 per cent level for manuring.

Comparing the predicted versus the actual outcomes can test the explanatory power of the above analysis.[5] The predicted outcome is 'yes' in case the predicted probability of adoption is 0.5 or more, and 'no' otherwise. Table 3.7 shows the cross-tabulation of the predicted and actual outcomes for terracing, manuring and fertilizer respectively ('Full model'). For comparison, we also report the same statistics for a model with village dummies only.

The table shows that the full model gives correct predictions in 78, 75 and 71 per cent of the cases for terracing, manuring and fertilizing respectively. However, a model with village dummies only, predicts 74, 64 and 61 per cent of the cases correctly. This is a sobering result – although the full model includes many variables at the plot and household level and many of these variables have a large and significant impact, a model with village dummies only performs relatively well. Of course, one possible explanation is that the decision to adopt depends strongly on unobserved idiosyncratic factors at the plot and household level that are uncorrelated with observable factors ('noise'). In that case one cannot expect major improvements in predictive power with changes in the specification of the adoption model.

An alternative explanation for the weak explanatory power of the model is that it omits crucial factors that affect the decision to adopt and that are also observable, in principle. In Chapter 2 of this volume it was found that terracing at the village level does not take place continuously, driven by constant pressure of drivers of adoption, but in short bursts and peaks, interspaced with period of very slow adoption. The time dimension, the dynamic character of the adoption process, seems to be very important at that level and this is totally lacking in the above analysis. We feel there should be a link between this dynamic process of adoption at the village level and the adoption process itself at the lower level of the household and farm. This relationship is explored in the following part.

---

5    The $R^2$ is not available for the logit model.

**Table 3.6    Logit regressions of the probability of terracing, manuring and fertilizing. Marginal effects evaluated at the mean (p-values in parentheses)**

| Variables | Terracing | Manuring | Fertilizing |
|---|---|---|---|
| Plot characteristics | | | |
| Distance plot to home (km) | -0.028 | -0.151*** | -0.017 |
| | (0.502) | (0.000) | (0.144) |
| Plot size (acres) | -0.007 | 0.004 | -0.001 |
| | (0.517) | (0.780) | (0.747) |
| Tenure | | | |
| Private title deed (dummy) | 0.161*** | 0.071 | -0.002 |
| | (0.007) | (0.493) | (0.979) |
| Still obtaining title deed/demarcated (dummy) | 0.285*** | 0.034 | 0.022 |
| | (0.000) | (0.778) | (0.809) |
| Slope | | | |
| Lower slope | 0.105 | 0.282*** | -0.051 |
| | (0.141) | (0.003) | (0.558) |
| Medium slope | 0.083 | 0.287*** | 0.008 |
| | (0.234) | (0.004) | (0.918) |
| Upper slope | 0.101 | 0.384*** | -0.021 |
| | (0.214) | (0.000) | (0.822) |
| Soil type | | | |
| Sandy/sandy/loam | 0.069 | 0.255** | 0.135 |
| | (0.405) | (0.026) | (0.139) |
| Loamy/silt | 0.009 | 0.328*** | 0.428* |
| | (0.913) | (0.003) | (0.084) |
| Household characteristics | | | |
| Household size | -0.020* | 0.017 | 0.016 |
| | (0.069) | (0.343) | (0.203) |
| Age composition (proportion) | | | |
| 18–39 years | -0.176 | -0.257 | -0.403 |
| | (0.213) | (0.305) | (0.189) |
| 40–59 years | -0.588*** | -0.149 | 0.100 |
| | (0.001) | (0.666) | (0.776) |
| 60 or more years | 0.179 | -0.464 | -0.456 |
| | (0.390) | (0.244) | (0.154) |
| Highest education in household (dummy) | | | |
| Secondary school | 0.284*** | 0.045 | -0.008 |
| | (0.003) | (0.807) | (0.966) |
| College | 0.314*** | 0.202 | 0.375** |
| | (0.003) | (0.288) | (0.018) |
| Off-farm or self-employment (proportion) | -0.096 | -0.020 | 0.268** |
| | (0.238) | (0.874) | (0.016) |
| Total land owned or used (acres) | -0.003 | -0.003 | 0.002 |
| | (0.489) | (0.529) | (0.530) |
| Cattle and smallstock (TLU) | 0.015** | 0.005 | -0.003 |
| | (0.022) | (0.762) | (0.547) |
| No information on livestock (dummy) | -0.058 | -0.405*** | -0.043 |
| | (0.487) | (0.000) | (0.680) |
| Village characteristics included | Yes | Yes | Yes |

*Source*: Authors' analysis based on NWO survey data. Note:  coefficients that are significant at a significance level of 1%, 5% and 10% are reported with *, **, and *** respectively. Standard errors are corrected for clustering at the household level.

**Table 3.7     Predicted versus actual adoption outcomes**

|  | Full model | | | Only village dummies | | |
|---|---|---|---|---|---|---|
| Terracing | Predicted | | | Predicted | | |
| Actual | No | Yes | Total | No | Yes | Total |
| No | 41 | 61 | 102 | 16 | 86 | 102 |
| Yes | 20 | 247 | 267 | 10 | 257 | 267 |
| Total | 61 | 308 | 369 | 26 | 343 | 369 |
| Per cent correct | 67 | 80 | 78 | 62 | 75 | 74 |
| Manuring | Predicted | | | Predicted | | |
| Actual | No | Yes | Total | No | Yes | Total |
| No | 117 | 56 | 173 | 97 | 76 | 173 |
| Yes | 36 | 160 | 196 | 58 | 138 | 196 |
| Total | 153 | 216 | 369 | 155 | 214 | 369 |
| Per cent correct | 76 | 74 | 75 | 63 | 64 | 64 |
| Fertilizing | Predicted | | | Predicted | | |
| Actual | No | Yes | Total | No | Yes | Total |
| No | 211 | 80 | 291 | 181 | 110 | 291 |
| Yes | 26 | 52 | 78 | 34 | 44 | 78 |
| Total | 237 | 132 | 369 | 215 | 154 | 369 |
| Per cent correct | 89 | 39 | 71 | 84 | 29 | 61 |

## The Dynamics of Terrace Construction at the Plot Level

The multilevel model in the previous section shows the importance of household- and plot-level characteristics but still explains only a small part of the adoption pattern in terracing, manuring and fertilizing. This lack of explanatory power could be the result of the static nature of the model. Indeed, the dynamic analysis of terrace adoption in Chapter 2 suggests that village-level characteristics can explain a major part of the village-level adoption pattern over time. For instance it was found that if the travel time to Nairobi had been double, the cumulative rate of adoption would be at most 32 per cent in 1996, as opposed to the observed 55 per cent. Also the coffee price was found to have a significant impact on adoption – if coffee prices had stayed at their 1966 (real) price level, then the cumulative adoption rate would have been 39 per cent.

The analysis of variance of adoption suggests, however, that much of the variation in terrace adoption is found at the household and plot level rather than the village level. Therefore, given that Chapter 2 has focused on important village-level factors explaining village-level adoption patterns, the issue arises which household- and/or plot-level characteristics may be important factors for understanding the dynamics of terrace adoption at the *household* and *plot* level.

In this section we will focus on one particular household/plot-level characteristic, namely ownership changes, for understanding the dynamics of terrace adoption at the household and plot level. Because we focus on the dynamics of adoption, we limit the analysis to terracing, as we do not have (retrospective) information on the use of manuring and use of artificial fertilizer in the past. The analysis will show that any analysis of terrace adoption should take into account that a significant part of the terracing takes place within a few years after the change in ownership. This finding holds across different decades and also for ownership changes due to inheritances as well as land purchases. In the subsequent section we will discuss whether and how these findings can be understood within the existing literature on soil and water conservation adoption.

### *A survival analysis of terrace construction in Machakos and Kitui, 1940–98*

A dynamic analysis of terrace construction involves a 'duration' or 'transition' or 'survival' analysis (e.g. Cameron and Trivedi 2005, Chapter 17). Each plot is in two possible 'states', namely 'not terraced' and 'terraced' and undergoes a 'transition' if it moves from one state to another state. We are interested in the transition from the state 'not terraced' to the state 'terraced', and the time ('duration' or 'spell' length) it takes before a plot is terraced. The literature on duration or transition or survival analysis is quite large and has been applied in many fields, such as engineering, biostatistics and social sciences. Also in economics there are many applications of survival analysis, for instance on the average length of an unemployment spell or the duration of strikes. There are also a limited number of applications of survival analysis in the literature on soil and water conservation, such as Dzuda (2001), Burton et al. (2003) and D'Emden et al. (2006). These studies do suggest that adoption is a dynamic process and therefore that time-dependent economic and environmental variables do affect the adoption decision. Of particular relevance is the study of Dzuda (2001), who noted that adoption is more likely in the first five years of the 'management life' of the farmer, and we will return to this study later on.

In order to do a more formal analysis, we introduce a number of basic concepts as used in survival analysis. Let the duration that a plot is not terraced be given by $T$. A central concept in survival analysis is the 'survivor function' $\hat{S}(t)$, which gives the probability that a plot has not been terraced by time $t$:

$$S(t) = \Pr[T > t]$$

The survivor function is monotonically declining from one to zero. If all plots will be terraced eventually, then $S(\infty) = 0$. If not all plots will be terraced eventually, then $S(\infty) > 0$. It can also be shown that the area under the survivor function gives the mean duration of a plot before it is terraced.

Because we can only observe the current and past state of a plot, the data are 'right-censored', in the sense that we observe the state of a plot from time 0 until the censoring time 1998 (the final year before the survey). Some plots will be terraced by the time of the survey ('completed spells'), but others plots are still not

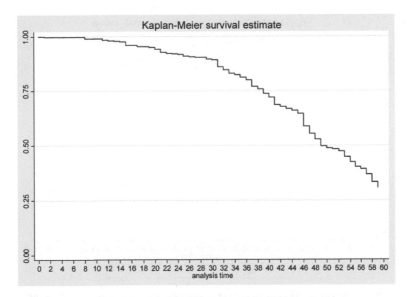

**Figure 3.1    Kaplan-Meier survival estimates for all plots, 1939–98**

terraced and all we know is that they may be terraced at some point in the future ('censored spells').

The survivor function cannot be observed but can be estimated from duration data. If there is no censoring, the obvious estimator of the survivor function is given by the number of plots that have not been terraced yet by *t,* divided by the total number of plots. However, with right-censoring, this will give a biased estimate of the survivor function because we no longer can observe all transitions. The Kaplan-Meier estimator of the survivor function provides a nonparametric and consistent estimate of the underlying survivor function in the presence of right-censoring. However, it assumes that the censoring process is independent of the terracing process and we will discuss this assumption at a later stage.

For each plot we know the time of first terracing (if any), and Figure 3.1 shows the estimated Kaplan-Meier estimator that a plot has not been terraced so far. We choose as starting point 1939 because the first plots were terraced in 1940. Interestingly, the survivor function is concave, reflecting the fact that the probability that a plot is not yet terraced increasingly falls over time. The concavity of the survivor function is mirrored in the convexity of the cumulative distribution of plots terraced as reported in Figure 2.2 in Chapter 2 of this volume. This is because without censoring the survivor function equals one minus the empirical cumulative distribution function (Cameron and Trivedi 2005, p. 582), and censoring takes only place at t=59 in the figure.

In Figure 3.2 we distinguish between two types of plots, namely plots that are owned by the farmer who also owns the plot at the time of the survey (1999),

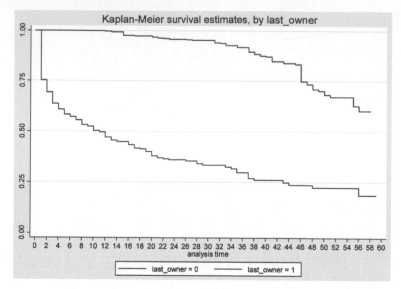

**Figure 3.2.	Kaplan-Meier survival estimates by acquisition, 1939–98**

and plots that are owned by a previous owner. This introduces additional right-censoring in the data, because a plot might change ownership from a previous to a current owner before 1999. Figure 3.2 shows the Kaplan-Meier estimates for the survivor functions for both types of plots. We note that the time is different for both types of plot. For plots under previous ownership it is the time since 1939, while for plots under current ownership it is the time since the ownership change. There are two remarkable findings. First, plots under current ownership are much more likely to have been terraced at any point in time. Second, the survivor function for plots under previous ownership is still concave, while the survivor function for plots under current ownership is *convex*. This convexity is particularly strong in the first years after ownership but appears to continue afterwards as well. This suggests that the probability that a plot is still not terraced does not increasingly fall over time and that much terracing takes place in the first years after ownership change.

One may argue that the above findings reflect a time effect, in the sense that in recent decades the plots tend to be under current ownership, while in past decades plots were predominantly owned by previous owners. We therefore also estimated the survivor function for the last four decades separately, comparing plots under different ownership at the same point in time. In this case the time is the same for both types of plots, namely the time since the start of the decade.

The figures show that plots under current ownership are still much more likely to have been terraced at any point in time than plots under previous ownership, even if we control for the time period. This suggests the existence of a cohort effect, in the sense that current owners tend to be younger than previous owners.

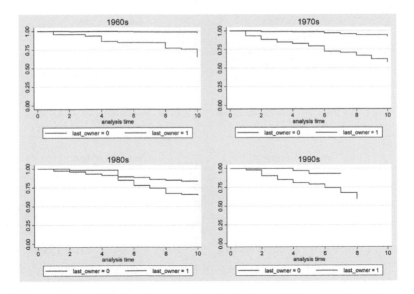

**Figure 3.3     Kaplan-Meier survival estimates by decade and ownership, 1939–98**

At the same time a comparison of the survivor functions across the decades suggests that there is a time effect as well, especially for the plots under previous ownership. The survivor function is extremely flat in the 1960s and also rather flat in the 1970s for plots under previous ownership, but it becomes much steeper during the 1980s and 1990s. This suggests that, over time, the plots under previous ownership also have an increasing probability of being terraced. This may indicate a neighbourhood effect, where previous owners less likely to adopt terracing are now also switching, in an environment with increasing numbers of new owners who are quickly terracing around them.

However, the most remarkable finding in the figure is something that is not there any more – namely the convexity of the survivor function for plots under current ownership. While the survivor function for this group of plots was clearly convex in Figure 3.2, it is no longer convex once we control for the time effect. This suggests that the convexity of the survivor function is somehow related to the adoption pattern around the time of acquisition. In Figure 3.4 we therefore redo the survival analysis of Figure 3.2 but controlling for the time effect. The survivor function for plots under current ownership is convex for each of the decades, and the reason is found in the significant activity in terracing right after ownership change. In particular, about 22–26 per cent of the plots are terraced within one

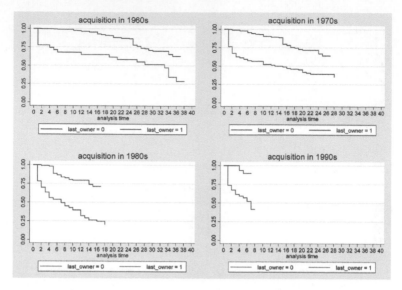

**Figure 3.4**    **Kaplan-Meier survival estimates by decade and acquisition, 1939–98**

year of the ownership change,[6] 29–46 per cent are terraced within five years,[7] and for the 1980s and 1990s 58–60 per cent are terraced within ten years.[8] The corresponding percentages for plots that are under previous ownership are 0 per cent for the first year, 1–12 per cent within five years[9] and 2–21 per cent within ten years.[10]

We have also checked whether the survivor function is different for plots that change ownership through inheritance and plots that change ownership through purchases, but no obvious differences were found. This suggests that the burst in terracing cannot be understood in terms of credit market imperfections, as plots that are purchased are not more likely to be terraced compared to plots that are inherited. In the following section we therefore turn to the literature to develop an understanding of the importance of ownership changes for soil and water conservation adoption.

As a final point, we noted that the Kaplan-Meier survival estimates assume that the censoring process is independent of the terracing process. In the above analysis this implies that the timing of transfer of ownership does not affect the subsequent timing of adoption. In other words, the shape of the survivor function for plots

---

6   By decade: 1960s: 23%, 1970s: 24%, 1980s: 22% and 1990s (till 1998): 26%.
7   By decade: 1960s: 29%, 1970s: 41%, 1980s: 46% and 1990s (till 1998): 43%.
8   By decade: 1980s: 60% and 1990s (till 1998): 58%.
9   By decade: 1960s: 1%, 1970s: 2%, 1980s: 12% and 1990s (till 1998): 10%.
10   By decade: 1960s: 2%, 1970s: 10%, 1980s: 21% and 1990s (till 1997): 10%.

under current ownership does not depend on the timing of ownership change. Of course, the survivor function may depend on factors such as time period and the characteristics of the owner, as Figure 3.4 suggests, but not on the *timing* of ownership change.

This appears to be a reasonable assumption but its validity will ultimately depend on our understanding of why ownership change would matter for the decision to adopt soil and water conservation techniques at all. In the next section we will explore the literature in search of possible theoretical explanations for this puzzling finding.

*Ownership changes and innovation, a puzzle for the literature*

The above time-based analysis of terracing at the plot level shows that ownership change of plots is a very important predictor of terracing on these plots. This is an interesting combination of a plot- and household-level characteristic, not a characteristic of the plot or the owner of household separately. This seems to be a trigger in the adoption process, possibly together with other driving forces providing background pressure on land managers to adopt the innovation.

The literature on soil and water conservation techniques has so far ignored this relationship between plot ownership changes and innovation. The only exception we are aware of is a study by Dzuda (2001), focusing on the time it takes by farmers to adopt soil and water conservation techniques in Zimbabwe. This study applies duration analysis and finds that a farmer is more likely to adopt SWC in the first five years of his management life, which overlaps with our conclusion assuming that management and ownership are tightly related. However, the study does not develop nor discuss any theory on *why* farmers are more likely to adopt in the beginning of their management life.

The question is then how the decision to adopt innovation is linked with a change of ownership. We can think of two possible mechanisms. First, the change in ownership is related to higher-level developments such as some of the driving forces we found earlier which also affect adoption behaviour (Chapter 2). For example, maybe a drought or a sudden increase of prices of certain cash crops drives an owner into bankruptcy, or to the decision to hand over or sell his farm, but at the same time stimulates the new owner to adopt new technologies. In another context this mechanism has been observed as well, namely regarding changes in livelihood – Dietz (1987) shows that with every serious drought, a number of sons from a pastoral household shift from the drylands to the more humid highlands of West Pokot, Kenya, to acquire land and start farming. But the literature remains extremely patchy in this respect ignoring the innovative consequences of drought, cash crop price rises and other drivers on farm management. Usually, the opposite is studied: how can innovation make farmers more resilient against drought and degradation.

Within such an analysis a timing-determinant variable (change of ownership) at the plot–household (farmer) interaction level would be related with regional time-dependent variables (transaction costs, population density, drought and price of coffee) as well as with farmer adoption behaviour. A change in ownership may affect farmer adoption behaviour through different channels, and for this the analysis should look at the characteristics of the farmers selling/handing over their land, and of those who acquire it. Are these latter people local farmers extending their farm, or are they new to the area and bring along the technology we see being adopted? Or is it rather the inheritance process itself that causes children to divide the farm and rely on the more limited area of land available for each son of the previous owner? In that case, intensification may be necessary to allow increasing numbers of people to live off the land.

A second possible mechanism between the decision to adopt innovation and a change of ownership could be found in the bodies of literature that deal with the process of transfer of ownership of assets, be it a family-owned company for example or the purchase of such a business. This suggests that we may have to look into the literature on family business wealth transfers and investment strategies rather than into agricultural innovation and its determinant drivers in a more limited sense. For example, Churchill and Hatten (1997) have studied market and non-markets transfers of family businesses and developed a framework that allows the study of the succession process where changes in management style, strategy and control mechanisms are actually planned for and executed. If for example we could link up with the process of training a new generation, the intergenerational partnership and the power takeover that they describe, and see why it is that only in the last stage the decision is taken to innovate, we could be closer to a theory that allows insight into the actual process of and decision to adopting innovations. The fact that in our case both inherited and purchased plots are quickly conserved may point at a related process of knowledge creation outside of the home and an intergenerational power struggle within the family business that needs to be addressed.

# Chapter 4

# Impact of Transaction Costs and Risks on Terracing in Kenya

Samuel M. Mwakubo, Michael K. Bowen, Wilson.K. Yabann and Henry K. Maritim

## Introduction

About 80 per cent of the total land area in Kenya is marginal for rain-fed agriculture. These areas are faced with frequent droughts and food shortages, are ecologically vulnerable and receive irregular and low amounts of rainfall. They also face very serious problems of environmental degradation such as soil erosion and soil mining. Soil degradation, seen here as a reduction in the land's potential use, is thus increasingly being regarded as a major, perhaps the most threatening environmental problem in developing countries (Reardon and Vosti 1992). The main negative consequence of soil degradation is on-farm decline of crop production, without belittling the other effects such as siltation of waterways. In developing countries, where substantial numbers of people still depend directly on agricultural production, the effect on yields is often critical. Yields decline partly because essential nutrients and organic matter are lost. Eroded soil also suffers from moisture deficiency because the subsoil structure is generally blocky, hard and dense compared to topsoil (Walker 1982).

Nevertheless, marginal areas can be very productive if farmers make substantial investments on their land. Such investments include terracing, application of manure and planting of trees. These investments conserve water and soil nutrients at the farm level, though the effect is not always straightforward as grass strips and trees also consume water and thus reduce availability of water for crops.

Investments in soil conservation may be undertaken when sufficient returns are expected. The returns to these investments critically depend on what the household can do with the crops. These returns, in particular monetary returns, when crops are sold, are influenced by transaction costs of market exchanges which subsequently determine the level of access to input and output markets. As Shiferaw and Holden (1998) argue, negative returns to soil conservation may undermine households' incentives to invest in conservation technologies.

Land tenure is the system of rights and institutions that govern access to and use of land and other resources. The rights are derived from statutory and customary law, as well as from institutions of marriage, of power and control and of inheritance. Whether customary or statutory, tenure regimes are rarely static, and the evolutions of customary tenure as well as the impact of directed land reform constitute two major strands of land tenure research. Insecure tenure rights on land and the imperfect functioning of the land market tend to reduce incentives for small farmers to invest in that land. A farmer may face lower expected returns from soil conservation investment because of the probability of being evicted before realizing all the benefits.

Besides the problem of soil erosion, farmers in the semi-arid lands of Kenya face a number of risks. Sources of risk in crop production activities may result from various factors influencing the crop production process. A broad distinction can be made among four types of factors (Anderson et al. 1977). They include decision factors under the control of the farmer such as weeding, fixed exogenous factors beyond the control of producers such as soil type, uncertain exogenous factors not controlled by farmers such as states of nature and finally output variability due to differences in managerial ability.

This study therefore focused on the influence of transaction costs, land tenure and risks on investments in soil conservation on agricultural holdings of households. Some of the questions addressed were: Do areas having high levels of transaction costs have low soil and water conservation investments? Is having a title deed to land significantly different from that of traditional tenure arrangements? What sources and types of risks with their impacts are prevalent in semi-arid Kenya? What is the linkage between transaction costs, land tenure, risk assessments and other factors influencing resource use?

## Description of Variables Related to Transaction Costs and Tenure Arrangements

The data used in this study come from a survey of rural households in Machakos and Kitui districts in the 1999/2000 cropping season, described in detail in Chapters 2 and 3 of this volume. The description and measurement of the variables that were used in the model presented above are given in Table 4.1. In this study, the length of terrace per hectare is taken as a proxy for soil conservation investments, because terrace construction is the major soil conservation measure. Because of the heterogeneous nature of the crops grown such as maize, beans, bananas and citrus the market value was used to obtain the aggregate crop output from the whole farm. Two variables represented transaction costs. Search costs are the opportunity cost of time for a farmer to find a buyer while access costs are the transport costs per person to the major market in each district in Kenya shillings.

Because of the simultaneity of the investment phenomena and feed back effects, a total of five equations were estimated simultaneously as a system using three stage least squares (3SLS).

Under the hypothesis of a stable relationship between risk aversion attitudes, between unchanged transaction costs due to terracing, and between degree of tenure security and related level of adoption over time, this model can explain

**Table 4.1     Description and measurement of variables**

| Variable | Description | Measurement |
|---|---|---|
| SLOPE | Average slope of land parcels in the household | Simple scale: flat (1), medium slope (2), steep slope (3) |
| TENURE | Land tenure regime | Simple scale of increasing tenure security |
| LOC | Dummy for location of study site | Kitui(1), Machakos(0) |
| DISTHOME | Total Distance of crop fields to the homestead | Metres |
| CROP OUTPUT | Crop output | in value terms |
| SEARCHC | Search costs for finding a buyer of farm produce | Kenya shillings |
| EDUC | Education level of household head | Simple scale: 0 no education, 1 primary, 2 secondary, etc. |
| WEALTH | Wealth of the household | Number of rooms in main house |
| SEX | Sex of household head | Male (1), female (0) |
| FARMOR | Degree of farm orientation | Fraction of off-farm income |
| SHELPG | Self help group | Member (1), otherwise (0) |
| HHSIZE | Household size | Number of persons |
| FARMSPC | Farm size per capita | Hectares per person |
| INCOME | Household income | Kenya shillings |
| AGE | Age of household head | Number of years |
| ACCESSCOSTS | Cost of access to the market | Kenya shillings |
| ERODE | Farm eroded or not | Eroded (1), otherwise (0) |
| TERRACE | Length of terrace | Metres per hectare |
| LABOUR | Labour use | Man-days per hectare |
| FERTILIZER | Fertilizer use | Kilograms per hectare |
| MANURE | Manure use | Kilograms per hectare |

the influence of the various factors on terrace investments. The production function relating output per acre to inputs (terracing, manure, fertilizer, labour) is the central equation, and the other four equations relate to each input. Terracing, represented by length of terrace, is seen as a function of *plot characteristics* such as slope, tenure, distance to the homestead, extent of farm erosion, and *household characteristics* such as marketing search costs, education, wealth, sex of the head, extent of off-farm employment, location. The other input variables are related to mostly the same variables.

The dependent variables were terrace length per hectare, crop output per hectare, labour use per hectare, manure use per hectare, and fertilizer use per hectare. Since manure, fertilizer use and terracing intensity are censored variables, the Heckman Two-Stage estimating procedure was used to accomplish this model (Maddala 1983). In the first stage, the inverse Mills ratio was computed and then added as an explanatory variable in the system equation estimation in the second stage. A number of independent variables such as education level, membership of a self-help group, slope and tenure, among others, were used to assess their influence on the dependent variables.

## Results and Discussions

Of all the individual fields sampled, 68.9 per cent had been terraced, while 31.8 per cent had not. This suggests that soil conservation is generally taken seriously in the two districts. Discussions with farmers and soil conservation officers reveal that terracing is carried out irrespective of whether the land is flat or not.

The difference comes in the terracing intensity. On a flat parcel of land, terraces are constructed 25–30 metres apart implying that the total length per hectare would be much smaller.

The mode of acquisition remains mostly by inheritance (70 per cent of all the fields) followed by purchasing (about 27 per cent). This shows that the land market is not very active in the study area, a phenomenon associated with market imperfection. The results also suggest little mobility of people from and to the study villages. Further, even if people migrate, some members of the family are left behind thus continuing holding their ancestral land. The results are corroborated by studies both in Kenya and elsewhere in Africa that land markets are inactive (André and Platteau 1998, pp. 18–19). Supply considerations largely explain why land sale markets are thin in Sub-Saharan Africa, 'distress sales' being an important reason for supply of land (Bardhan 1984). Landholders are typically reluctant to sell their land, even when they get an employment outside the agricultural sector and they reside in town. This is because land continues to be perceived as a crucial asset for the present and/or future subsistence of the family, all the more so as it is a secure form of holding wealth and a good hedge against inflation. Such security considerations often underlie the apparent persistence of indigenous control of land transfers even when they are duly registered.

With tenure, we observed that individual parcels having title deeds and those in the process of obtaining title deeds are the most predominant (38.8 per cent and 49.9 per cent respectively). This shows considerable interest in improved or increased tenure security by farmers. Table 4.2 below shows the tenure regimes for the fields of sampled farmers vis-à-vis whether they were terraced or not. If we combine the first two (essentially the same group with differences being in time), the percentage of plots terraced in the combined regime is 64.1 per cent. This shows clearly the importance of obtaining title deeds (titling) on investment in sustainable land use. As Tiffen et al. (1996) argue, secure land tenure is important to farmers' willingness to invest in land improvement, most particularly in long term measures such as soil and water conservation. As Ervin (1986) and Wachter (1992) argue, insecure property rights dissuade farmers from undertaking long-

**Table 4.2     Land tenure and whether fields are terraced or not in Machakos and Kitui districts, 2000**

| Tenure regime | Terraced | Not terraced | Total |
|---|---|---|---|
| Private title deed | 118 (28%) | 42 (10%) | 160 (38%) |
| Still obtaining title deed | 151 (36%) | 35 (8%) | 186 (44%) |
| Traditional private rights | 43 (10%) | 21 (5%) | 64 (15%) |
| Communal rights | 5 (1%) | 0 (0%) | 5 (1%) |
| Squatter | 0 (0%) | 1 (0%) | 1 (0%) |
| Rented in | 0 (0%) | 3 (1%) | 3 (1%) |
| Rented out | 0 (0%) | 1 (0%) | 1 (0%) |

*Source*: Field Survey, 2000

term investments, such as investments in soil conservation, because they may not be able to reap the benefits of such investments.

The study also shows that 73.3 per cent of the households were male headed, 14.8 per cent female operated and 11.9 per cent female headed. Female-operated households are households that are headed by men who do not reside in their homes. In this scenario, women operate the households though headed by men. Of the sampled households, 66.7 per cent of the households had members involved in self-help group activities. The predominance of collective action suggests that the transition to sustainable farming may depend on the ability of the community to cooperate, learning and coping mechanisms and social norms about good farming. Where collective action is predominant, transaction costs generally tend to be lower.

*Econometric results*

The 3SLS analysis was done at the household level. Table 4.3 shows that investment in terraces, here measured by the log of metres per hectare, is significantly influenced by slope, tenure, location, distance to fields, education, wealth, degree of farm orientation, membership in self-help groups, household size, farm size per capita, household income, age and transaction costs.

The slope of the farms appears to have a significant negative direct effect on soil conservation investments while the indirect effect is positive and significant through manure use. Steeper parcels of land are more susceptible to erosion and it is expected that steepness discourages the use of chemical and organic inputs because of run-off. However, the results show that manure and fertilizer application increase with slope suggesting that farmers sometimes increase their use even on steep slopes in order to maintain production levels. The negative correlation shows that the relationship between conservation investments and field slope is rather complex. Farmers invest most heavily on slopes of medium steepness (those steep enough to need conservation investments), but not so steep as to discourage investments, as their maintenance is very costly (Clay et al. 1996). In our case the net effect of slope is negative, due to the difficulty of manoeuvring

**Table 4.3**    **3SLS regression results for determinants of inputs and output in Machakos and Kitui districts, Kenya, 2000**

| Equations | 1 (Terrace) | 2 (Manure) | 3 (Fertiliser) | 4 (Labour) | 5 (Crop output) |
|---|---|---|---|---|---|
| *ln SLOPE* | -0.235 | 0.639 | | | |
| | (-1.742)** | (4.193)*** | | | |
| ln *TENURE* | 0.754 | | 1.401 | | |
| | (1.450)* | | (2.776)*** | | |
| *LOCATION* | -0.936 | | -0.986 | | 0.341 |
| | (-5.659)*** | | (-6.202)*** | | (0.959) |
| ln *DISTHOME* | -0.800E-01 | | 0.575E-01 | | |
| | (-2.639)*** | | (1.962)** | | |
| ln *SEARCHC* | | -0.461 | | | |
| | | (-7.593)*** | | | |
| ln *EDUC*+1 | 0.311 | -1.059 | 1.096 | -0.288 | |
| | (1.558)* | (-4.730)*** | (5.706)*** | (-2.211)** | |
| ln *WEALTH* | 0.552 | 0.526 | 0.496E-01 | 0.136 | |
| | (3.667)*** | (3.101)*** | (0.341) | (1.385)* | |
| SEX | | 0.994 | -0.540 | | |
| | | (3.461)*** | (-2.186)** | | |
| ln *FARMOR* | 0.326 | 0.868 | | 0.597 | |
| | (2.863)*** | (6.784)*** | | (7.711)*** | |
| SHELPG | 0.202 | 0.329 | | | |
| | (1.387)* | (2.005)** | | | |
| ln *HHSIZE* | -0.525 | 0.402 | -0.482 | -0.619 | |
| | (-2.675)*** | (1.828)** | (-2.554)*** | (-4.889)*** | |
| ln *FARMSPC* | -0.554 | | | -0.793 | |
| | (-5.159)*** | | | (-9.033)*** | |
| ln *INCOME* | 0.554 | 0.814 | 0.133 | 0.540 | |
| | (6.484)*** | (8.498)*** | (1.621)* | (9.478)*** | |
| ln *AGE* | 1.178 | -1.362 | 1.057 | | |
| | (3.704)*** | (-3.831)*** | (3.455)*** | | |
| ln *ACCESSCOST* | -0.567 | -0.960 | | | |
| | (-2.664)*** | (-4.009)*** | | | |
| *ERODE* | -0.352 | -0.935 | -0.401 | | |
| | (-2.470)*** | (-5.821)*** | (-2.908)*** | | |
| ln *TERRACE* | | | | | -0.231E-03 |
| | | | | | (-0.786E-02) |
| ln *LABOUR* | | | | | 1.056 |
| | | | | | (6.791)*** |
| ln *FERTILIZER* | | | | | 0.596E-01 |
| | | | | | (1.053) |
| ln *MANURE* | | | | | 0.206E-01 |
| | | | | | (0.868) |
| IMR | 3.402 | 3.364 | 1.704 | | |
| | (30.25)*** | (31.76)*** | (15.35)*** | | |
| *CONSTANT* | -3.926 | | -6.433 | -2.366 | 4.490 |
| | (-1.728)** | | (-2.947)*** | (-1.640)* | (5.869)*** |
| N | 148 | 148 | 148 | 148 | 148 |

\* significant at P<0.10, ** significant at P<0.05, *** significant at P<0.01

Figures in parentheses are t-statistics that the probabilities of respective coefficients are zero

*Source*: Field Survey 2000

draft animals, and the likelihood of terrace walls collapsing as slope increases further. Econometric evidence from Rwanda and the Philippines show that the net benefits of soil conservation are highest on fields of medium steepness (Clay et al. 1995; Templeton 1994). Soils on steep slopes are light and thin making them prone to erosion; they keep yields low and have lower long-term returns to investments. Thus a spiral of low production and low investment is set into motion on these marginal lands.

Security of tenure positively influences the use of fertilizer. Perceived security of tenure is important when investing in soil quality and conservation measures because of sunken costs in physical infrastructure (terraces, ditches), in long-term productivity and in knowledge acquisition. Given the long gestation period of terraces and the build-up of soil quality and fertility (Pagiola 1994), land titling and other mechanisms of increasing security of access to land are thus important for soil conservation investments (De la Brière 1999). Tenure reflects what Feder et al. (1985) term degree of 'confidence in the long term'. Tenure status also influences risk behaviour. We expect farmers to make fewer longer-term land improvements such as terraces on holdings that are rented in. These holdings have short-term use rights, and as such put long-term investments at the risk of re-appropriation by the owner (Migot-Adhola et al. 1990; Place and Hazell 1993).

Location of the households (with a value of 1 if in Kitui) is negative and significant. It is also significant indirectly through fertilizer use. There is a locational factor that favours Machakos with regard to fertilizer use, the proximity to Nairobi which makes fertilizers cheap due to lower transport costs and diffusion of knowledge much higher compared to Kitui District. The results also show that Machakos has significantly higher level of soil conservation investments compared to Kitui, and this confirms earlier results (Zaal and Oostendorp, Chapter 2 in this volume).

Distance to fields has a negative and significant direct effect on soil conservation investments. More distant parcels of land are difficult to supervise, control and monitor. Moreover, such parcels face high effective input costs at all levels of input use, the more so as means of transport, including donkeys, are in limited supply. Moreover, farmers fear theft of produce from parcels of land farther away, which is another form of transaction costs. Parcels of land closer to the homestead are better managed and are also likely to have been acquired much earlier. This is because one has to settle first before thinking of expanding or acquiring other parcels of land. Interestingly, this relates to the discussion by Oostendorp and Zaal (Chapter 3 in this volume), in which it is indicated that plots that are obtained (later) are being terraced to a large degree within three to five years; this may be comprehended as a form of land management which has been already implemented on the older plots closer to the house. In that case, it is a matter of incorporating a plot in the farming system, including the necessary land management.

Another compelling factor is social recognition or prestige linked with being a 'good' farmer. Indirectly, the effect of distance to the crop fields is also negative but weak through labour and manure use. However, with fertilizer use, the results are surprising as the correlation is positive and significant. But as Grabowski

(1990) argues, and 93 per cent of our sample farmers confirm, plot scattering or dispersion takes advantage of micro-climate variations and reduces the possibilities that a farmer's full range of crops will be lost to pests or weather problems. This suggests that sometimes risk considerations may conflict with a rational response to increased distance to the crop fields. Some studies view fertilizer use in a high-risk environment basically as a risk-reducing device (Antle and Crissman 1990; Van den Berg 2002). Besides, chemical inputs such as fertilizer are easier to transport compared to manure, which is bulky.

The level of education of heads of households is equally important. Its direct influence on soil conservation investments is positive and significant. Education, which is a proxy for information flow, may overcome many characteristics of farmers which act as obstacles to soil conservation. The indirect effect of education is also strong in farmers' manure, fertilizer and labour use. Manure use is often associated with the less educated. Thus the coefficient for the education level of principle household member is negative and significant. The results also show that fertilizer use is significantly influenced by the education level of the principle household member. Education leads to better resource allocation and is a form of human capital (Shultz 1964; Welch 1978; Pudasiani 1983; Idachaba 1994), besides improving the farmer's management capabilities (Gould et al. 1989). Other studies have also found a positive association between education and adoption of conservation technology (Earle et al. 1979; Ervin and Ervin 1982).

Wealth is positive and directly significant. The indirect effect of wealth is also significant in terms of manure and labour use. With fertilizer, the sign is as expected although not significant. Our considered inference is that wealth is conspicuous for there is prestige in having well-laid-down terraces, as it is what constitutes 'a good farmer'. The results also point to financial constraints resulting from imperfect credit markets (Shiferaw and Holden 1996; Pender and Kerr 1996). When credit markets are imperfect, which state is the norm in Kenya, wealth may ease investment cash constraints and also provide a sense of security (lower risk) to the household, which may enhance the adoption of conservation and the increased use of manure, fertilizer and labour. Wealthier households may have greater access to capital and thus increased possibilities for buying land and implementing soil conservation investments at the same time (see Oostendorp and Zaal in this volume).

The level of one's assets (wealth) also affects the degree to which one discounts possible future gains. Those who posses a higher quantity and quality of endowments will place a higher value on the medium and long-term benefits produced by investment technologies. This is because they are less constrained by food insecurity and risks, which undermine the ability to meet basic needs in low-wealth households. Poverty therefore may lead to high rates of time preferences and an inability to forgo immediate consumption to improve the future productivity of environmental resources (Holden et al. 1998). Moreover, the poor depend more on annual crops, which typically degrade soils more than other crops. Social structures and power distribution furthermore bias technologies and the flow of technical information in favour of the wealthy, thus shaping adoption outcomes (Grabowski 1990). Moreover, even the extension service has often favoured the wealthy (Knox et al. 2002).

In terms of the sex of the principle household member, we find that it is only positive and significant on manure use. The variable has a negative effect on fertilizer use. All other effects, whether direct or not, are weak. Discussions with extension staff reveal that women do not own livestock from which a substantial proportion of manure used is sourced. Female-headed households, however, use more fertilizer. This suggests that women, unable to access manure, apply more fertilizer on their farms. Moreover, fertilizer is easier to transport since it is not bulky. Our discussions with extension staff also revealed that women are exposed to extension services more than men. This is because they are often members of women groups. The extension service in Kenya often uses groups to pass agricultural information to farmers.

The fraction of household income provided by the farm itself is positive and significant both directly and indirectly. It is only with fertilizer use that this effect is negative although not significant. We can infer that if income from the farm is the predominant source of household income, there will be more concerted efforts made to either maintain or improve this source of income. Expectations about future income and household welfare depend on the farmer's planning horizon and discount rate (Solow 1974). The negative effect of degree of farm orientation with fertilizer use suggests two things. First, the presence of imperfect credit markets, despite the existence of merry-go-rounds; and secondly, that it is off-farm income that encourages fertilizer use. As Pender and Kerr (1996) argue, a negative coefficient of the share of income earned from farming (which suggests that off-farm income has a positive effect on fertilizer use) is due to financial constraints.

We also find that membership of self-help groups is important directly and as it affects manure use. These groups are based on the principle of reciprocity and are helpful when the option of hiring in labour is limited by liquidity constraints, especially when farmers cannot borrow against their future income (Oostendorp 1998). As Lindgren (1988) argues, most of the terraces have been built by the farmers themselves or by self-help groups. Membership of a self-help group is a form of social capital and is instrumental in reducing transaction costs. Moreover, it is a form of peer pressure: the farmer feels a need to terrace in order to gain acceptance in the society. Self-help groups are non-market institutions and are more appropriate if adoption of better technologies has certain fixed costs, which can be met through group labour inputs. By serving as a risk sharing device, Knox et al. (2002) argue that collective action can alleviate food insecurity and other survival risks borne disproportionately by the poor to lower the degree of future discounting and therefore positively influence technology adoption. Thus societies with high social capital are likely to sustain high investments in sustainable farming.

Household size by itself also has a strong negative direct effect on soil conservation investments. The indirect effect is strong also through manure, fertilizer and labour usage. It seems, therefore, that large household size, which keep land/man ratios the same, discourages the investment in soil fertility maintenance. However, with manure use, the sign is positive. Manure is often bulky and thus more labour effort is expended in its transportation and eventual application on the farms, hence its positive sign. We are of the view that a large household size implies higher consumer-to-worker ratio, which further implies high dependency ratio. The correlation coefficient between household size and number of children is a significant 0.63. This suggests that the constraints imposed on the household by having more dependents materially affect labour availability. Moreover, the necessity to support a large family may shorten the planning horizon of the poor and hence discourage soil conservation (Shiferaw and Holden 1996).

Farm size per capita, also incorporating the household size variable in its denominator, is negative and significant. This is an indication of land scarcity as well as population pressure. We thus infer that land scarcity significantly increases terracing intensity. The implication of the signs of household size and land per person is that as land remains the same, and household size increases, the effects on terracing are negligible. If household size remains the same, but land increases, terracing falls. Hence farmers, when confronted with declines in production and enjoying no access to alternative agricultural land or migratory networks, may be forced to increase terracing intensity. It is also logically possible that those with large parcels of land have opportunities for crop rotation and leaving the land fallow instead of using costly soil conservation techniques. The net effects of the two variables for household size are positive on the use of manure, negative on fertilizer use and positive on labour use on the farm.

We also find that household income has a positive and significant direct effect on soil conservation investments. This finding is consistent with other studies in the past (Norris and Batie 1987; Sinden and King 1988). The indirect effect is also strong through manure, fertilizer, and labour use. The significance of household income in input use such as manure, fertilizer and labour suggests the existence of imperfect credit markets (Pender and Kerr 1996). Higher income enables farmers to purchase materials and equipment for soil conservation or hire labour. A further explanation is that there is a greater willingness to take risks with increasing income levels (Binswanger 1982; Antle 1987, 1989; Myers 1989) and thus higher soil conservation investments. Besides, risk is closely related to other factors such as wealth and education (Norris and Batie 1987).

The age of the principle household member is another important factor in soil conservation investments. The direct effect is positive and significant. Indirectly, it is strong in terms of manure and fertilizer use. The effect in terms of manure is negative: households with older heads use less manure than those who are younger. It is argued that older people have accumulated wealth in most cases and thus tend to use fertilizers more. Further, older people are less strong physically, implying less use of manure, which is often bulky. The argument advanced is that older people have more farming experience. A positive role suggests that farmers have to learn about the effects of a new technology before adopting it entirely.

This is important if risks of a new technology are unknown. About 73 per cent of the sampled farmers said that they try new things on small portions of land first. Older farmers also have accumulated more wealth (Nyang 1999) and simply may have had more time to finance terrace construction. A bivariate correlation of age of principle household member and number of rooms (a proxy for wealth) shows that it is positive (0.208) and significant at 0.01 level. But where land markets are absent and poverty is rampant, age raises the time preference (i.e. high discount rate) of the poor, which may lower the desire for further conservation (Shiferaw and Holden 1998).

Search costs, measured as the time needed to find a buyer, do not have significant direct effects on soil conservation. We assume these search costs are of a similar magnitude to those of searching for sellers of inputs like manure. The lack of significance may arise from the fact that search costs do not vary with the quantity of goods to be sold and can only be reduced over time. As time goes by, farmers are able to establish contacts or networks with buyers. It is also plausible that traders are within reach and are locals in which case search costs are very low. It is the indirect effect in terms of manure use that is strong (significant at one per cent). Manure can be sourced from neighbours and sometimes all the way from Kajiado District. Thus the higher the search costs, the lower the use of manure at the household level. The direct effect of transaction costs (access costs) on soil conservation investments is negative and significant, illustrating the disincentives of transaction costs. Indirectly, the strong effect is in terms of manure use. Production nowadays is not at a purely subsistence level, and so farmers do invest with the objectives of meeting household consumption requirements and of selling crops to obtain money and meet some household expenditure requirements. Higher transaction costs to the market implies lower returns to crop production because farmers are price-takers. Thus the higher the transaction costs faced, the lower the expected returns and, consequently, the lower the investment in soil conservation.

There is also likely to be an issue of enterprise selection versus food security. Farmers tend to select enterprises that minimize food insecurity; these happen to be low-priced commodities whereby the net returns for a profit maximizer appear ridiculous. Besides, about 76.8 per cent of the sampled households said that it was better to have a greater portion of land given over to food crops than to cash crops because the prices for cash crops fluctuate too much. Market transaction costs, such as transport and handling charges and time spent travelling to and from markets, create divergences between market and farmgate prices leading to imperfect substitutability between domestic and market supplies of food. If, as Fafchamps (1992) argues, basic staples account for large shares in the total expenditures of rural households, then high transaction costs in food markets raise returns to food self-sufficiency. The inference is that farmers' behaviour is influenced by the expected profitability of any investment made. The returns to terrace construction are the crops grown after the investment has been undertaken. When the costs of access to the market are higher, the net returns to the farmers decrease significantly, thus reducing the incentive for further terracing. Prior to making any investments in soil conservation measures, it appears that farmers have a definite understanding of the transaction costs to the market. It implies that the decision to invest is arrived at after taking into account the transaction costs faced.

The strong indirect negative effect of transaction costs on manure use suggests that manure is often not obtained from the farmer's farm. It implies that a lot of manure comes from either neighbours or from a place a distance away from the households. Our main thesis is that the negative influence of transaction costs on manure use will cause a consequently lower output in crop yields, which in turn will reduce further incentives for soil conservation investments through the feedback behavioural effects. The correlation with fertilizer, though not significant, has the expected sign. With labour use, the correlation is unexpected (positive), though not significant. We are of the view that as a response to high transaction costs, farmers either use family labour or source labour nearby.

At this point it is worth considering the issue of reverse causation. We measured the dependent variable in terms of existing terracing. This is likely the result of many years of investments and maintenance; hence, their presence cannot simply be related to present-day variables as the household size now, income variables etc. Indeed, the reverse causation that the present availability of soil and water conservation favours the generation of farm income, or even helps farm household members stay on the farm, is a serious probability. Having terraced fields may also have led to higher wealth and more orientation toward farm work as the main occupation. The finding that older household heads show more terracing in their fields also must be seen in this light: they may have had more opportunities to terrace at some point of time (and maintained this until the present). Hence, many positive effects found in Table 4.3 could possibly be the result of reverse causation. In this case, the signs of the regression analysis should not be interpreted in a causal way, but as showing the association between dependent and 'explanatory' variables amidst the other variables. The market access variables, however, are truly exogenous, assuming that the location of the farm is fixed.

The potential bias on the estimated coefficients in the model is that the coefficients of the explanatory variables positively correlated with (past) investments, such as income, orientation etc., are likely to biased upward, while the coefficients of the market access variables are – therefore – more likely to be biased downward. True effects of market access can, therefore, be substantially stronger than suggested by the coefficients.

### Risk as a Deterrent to Investments in Terracing

Rainfall unreliability is an exogenous and major source of risk in the semi-arid lands of Kenya and investment in soil conservation is crucial if the land is to support the population. Due to rainfall unreliability, farming is a very risky business. The risk of crop failure in the semi-arid zones is high, with a significant reduction in crop yields (Flavian and Hoekstra 1990). Price variability in Kenya also became a major source of risk to farmers with the onset of structural adjustment programmes (SAPs) which have resulted in significant changes in economic policies since the early 1980s. The removal of guaranteed prices and subsidies has led to increased price variation for outputs as well as inputs.

Apart from changes in external economic circumstances, households are also faced with a changing institutional environment. The numerous development activities imposed upon farmers in Kenya also include the collapse (or near collapse) or privatization of cooperatives and marketing boards that helped stabilize prices that farmers received.

The risks that the farmers face can act as a deterrent to the investments on the land. Investments on the land such as terracing, cut-off drains, planting of trees, use of certified seeds and use of manure/fertilizers can have positive impact on crop yields. This section seeks to find out the farmers' behaviour in adoption of soil and water conservation measures in the presence of risk.

This is particularly true if institutions for spreading risks are poorly developed or non-existent as is the case in the area of study. The result is that production is severely constrained as investments are reduced and only proven traditional methods are employed.

It is also common to find that farm households do not adopt or partially adopt new technologies even when these technologies provide higher returns to land and labour than the traditional technologies. Considerations of risk and risk aversion explain such production decisions. The riskiness with investment will lead to under-investment and this is especially true if farmers are risk averse rather than risk neutral.

Background risks (Lipton 1979) are unpredictable and may have a major impact on the viability of the household unit. They may range from sickness of household members, withdrawing labour force from income-earning activities and requiring extra outlays for medicine or hospital bills, to death, theft, the unexpected wedding of children or the burning down of a house. To cover such risks households are usually dependent on village-level social insurance institutions such as self-help groups, and mandatory contributions during funerals and weddings. Individual households may keep livestock or save to meet such calamities.

Rural households in developing countries face a substantial risk of income variability (Kurosaki 1998). Risk considerations are more important for poor farmers in these regions because their income is low and formal insurance arrangements are seldom available. Once they encounter adversity, they may have no other means but to sell their valuable assets, as a result of which transient poverty might become permanent.

Differences in risk attitudes across individuals and over time may be caused by interpersonal variation in preferences, as is often assumed by economists, or by intrapersonal variation, as has been assumed in the mathematical psychology literature and in some econometric models (Coombs et al. 1970; McFadden 1983).

The type of enterprise undertaken may also influence the attitude towards risk. This study assumes that differences in risk attitudes are due to one or a combination of the two factors cited by the authors above.

Risk preferences structures of farm families can be evolutionary if learning is considered. It is also expected to depend on the composition of the family and on the stages in the life cycles of the members of the family (Ghodake and Ryan 1981). According to Anderson (1980) attitudes towards risk vary systematically with age, follow a general pattern of increasing risk aversion with age and tend to be matched similarly in marriage contracts.

The concept of aversion to downside risk asserts that individuals generally avoid situations which offer the potential for substantial gains but which also leave them even slightly vulnerable to losses below some critical level (Menezes et al. 1980). The overriding motive of avoiding a disaster is similar to the safety-first principle. Individuals averse to downside risk are usually expected to be decreasingly risk averse.

Better decisions in a risky world can be made if information is available (Hardaker et al. 1977). One of the impacts of accumulated information is that excessive optimism or pessimism are corrected. A farmer who is learning about a new technology may initially be pessimistic about how it will perform on his own farm. He may well have failed in the past by giving too much credence to the optimistic claims of agricultural scientists, farm advisers or rural sales representatives and so will heavily discount any new claims. Yet if he takes the initiative to find out more, perhaps trying out the new technology on a small scale if this is possible, his initial scepticism may be modified. His subjective distribution will tend to approach the true but unknown distribution as more information is accumulated.

A number of researchers have used the econometric approach to measure attitudes towards risk. They include Bigman (1996), Antle (1987), Coombs et al. (1970), McFadden (1983), Moscardi and de Janvry (1977) and Wolgin (1975). Antle (1987) showed that econometric risk attitude estimation is possible under less restrictive conditions. He applied his model in semi-arid rural India. The econometric method requires farm-specific production data such as output and input quantities, their prices and observable technological characteristics of the farms. The production data are assumed to be generated by farmers solving a single period maximization problem. Inputs are chosen conditional on information available before production begins. This assumption means that inputs are predetermined variables relative to output.

All the farms are assumed to produce using a similar technology, in the sense that any technological differences can be accounted for using observable variables. The stochastic technology can be represented by a joint probability distribution of outputs or by the corresponding distribution of revenue or profit.

The individual field data for production of a particular crop are desirable for the analysis of risk attitudes because they are likely to be more accurate and subject to less aggregation bias than whole crop or whole-farm data (Antle 1987). Here we closely follow the model used by Moscardi and de Janvry (1977), who used an econometric approach to determine attitudes towards risk under a safety-first approach. Risk is introduced in a model of economic decision-making as a safety-first rule. According to this rule, an important motivating force of the decision-maker in managing the productive resources that he controls and, in particular, in choosing among technological options is the security of generating returns large enough to cover subsistence needs. In their model, and under the assumptions they maintain, the degree of risk aversion manifested by individual peasants can be derived from observed behaviour. The assumptions include a power form for the production function, a linear form for the household's trade-off between expected income μ and its standard deviation (with parameter K showing the trade-off), and that the coefficient of variation θ of farm income is constant, so that changes in its mean are also reflected in its standard deviation. Under these assumptions, the first-order condition for maximization of the household's utility leads to:

$$K = \frac{1}{\theta}\left(1 - \frac{P_i X_i}{P\varphi_i \mu}\right)$$

Here, $\varphi_i$ is the elasticity of production (price P) of the $i^{th}$ input $X_i$ (price $P_i$). In this equation, the left hand side is the expected value of the marginal productivity of the $i^{th}$ input. On the right-hand side, the price of the $i^{th}$ input is compounded by the risk factor. The factor K is assumed to be a function of the socio-economic characteristics of the household. This equation then leads to an expression for K, that can be individually measured at the farm level

This is a measure of risk aversion that can easily be derived for each household from knowledge of the production function, the coefficient of variation of yield, the product and factor prices and the observed levels of factor use. Following Antle (1987), the production data are assumed to be generated by farmers solving a single-period maximization problem: inputs are chosen conditional on information available before production begins. This assumption means that inputs are predetermined variables relative to output.

Theoretically the variables below determine whether farmers will undertake investments on their farms. It is the hypothesis of this study that risk and attitude towards risk are important variables in explaining farmers' investment decisions. The model determines the impact of risk on land investments. A maximum likelihood estimation approach was used given the many proxy variables used in the study. Specifically, a Tobit model was employed in the regression analysis since it best fitted the data when compared to other models. The model can be described as follows:

Land Investments = f(Risk, Attitude toward Risk, Labour, Education Level, District, Sex, Member of Group, Age of the Head of Household).

Of these variables, the measure for attitude toward risk (the K derived above) turned out to be insignificant. As measure for the riskiness of the process itself the subjective estimates of the probability of crop failure was used. Such probability is quite high (60 per cent). This measure, indicated by risk in Table 4.4, shows a small and barely significant negative effect on the likelihood of having SWC.

Looking at the risk variable in the regression shows that the higher the risks, the less the investments and this is significant at the ten per cent level. Alternatively, this can be interpreted in terms of the impact of terracing on risk. Following this approach the results reveal that farms with less investment in terracing have higher variability in crop yields. The inference from this is that investment in terracing reduces risk. This has further implications for this paper given that farmers in such a semi-arid region are known to be risk averse (66.1 per cent of the farmers in this study were found to be risk averse; see Bowen 2005). Despite the aversion to risk, the farmers are rational and see a way of reducing risks by investing in terracing and therefore increase the probability of harvesting even when rains fail.

The results from Table 4.4 above also show that education has a positive impact on terracing, and is significant at 15 per cent level. This means that education is important in adoption of terracing as a soil and water conservation measure. Education affects farming in the following ways. The first one is through the worker effect where a farmer becomes more efficient in performing certain tasks. Second, farmers learn to choose optimal resource combinations. This could include choice of land investments and specifically terracing as one of the inputs. Education also influences the ability of the farmer to acquire and analyse available information on expected costs, returns, variability and innovations thus reducing time lags in adoption. Finally education improves the farmers' capacity to exploit new market opportunities. Education can also increase farmers' access to credit but at the same time increase the opportunity cost of farmer's time. It also provides farmers with information on conservation measures and the effect of soil erosion

**Table 4.4    Tobit regression results: impact of risk on terracing Machakos and Kitui Districts**

| Variable | Coefficient | t-statistic | Marginal effects |
|---|---|---|---|
| District: 1 if Machakos | 307.77 | 4.91*** | 0.031 |
| Education | 4.48 | 0.64 | 0.00045 |
| Risk | -0.52 | -1.68* | -0.0000052 |
| Sex | 45.69 | 0.66 | 0.0046 |
| Group: 1 if member | 56.42 | 0.98 | 0.0056 |
| Labour | 0.16 | 2.29** | 0.000016 |
| Age of head | 0.30 | 2.09** | 0.000030 |

*significant at $P<0.10$,  **significant at $P<0.05$, ***significant at $P<.01$

Log likelihood function = -1136.24, N=193

*Source*: Field Survey, 2000

on productivity. In the study districts all these effects of education are expected to come into play as household members acquire education.

The results further above show that heads of households who are members of a group invest more in terracing though this is not significant at the ten per cent level. It is also one of the variables with the largest magnitude in marginal effects meaning that increasing the number of households who are members of groups will have a large positive impact on terracing. The marginal effect of group membership in Machakos District is quite high. Households should therefore be encouraged to be group members since this would increase terrace construction, which would have the ultimate effect of increasing land productivity. This is consistent with the findings of Tiffen et al. (1994).


## Conclusions

Our main conclusion is that transaction costs both directly and indirectly reduces soil conservation investments. Despite inconsistencies in some instances, transaction costs (search or access costs) to the market are indeed important and negatively influence both terrace construction, and the use of manure and fertilizer. Subsequently, lower manure and fertilizer use leads to a reduction in soil conservation investments through negative feedback effects via aggregate crop output or yields.

The results further suggest that, in general, labour use increases with transaction costs. This appears to be a response by farming households in the face of subsistence needs and high transaction costs, and a failing labour market and other input and output markets. Most of this labour is basically family labour.

The results have also shown the importance of secure land tenure towards investments into sustainable land use. This illustrates the need for policies aimed at titling of land hand in hand with other relevant complementary policies if sustainable land use is to be achieved.

Third, despite the advantages of terracing, the presence of risk (60 per cent chance of crop failure arising out of weather risk alone) has a negative significant impact on terracing. Reducing these risks should be an important objective for both government and non-governmental organizations involved in agriculture. The reduction of these risks would increase land investments via incomes out of farming.

# Chapter 5

# The Productivity of Indigenous Soil and Water Conservation in Benin[1]

Esaïe Gandonou and Remco Oostendorp

## Introduction

There are still few studies to assess returns from indigenous mechanical and non-mechanical land conservation structures found in many areas of LDCs, particularly in Africa. Byiringiro and Reardon (1996) have examined the effects of soil conservation investments on farm productivity in Rwanda, and found that farms with greater investments in soil conservation have much better land productivity than other farms. Place and Hazell (1993), using data from Ghana, Kenya and Rwanda, were left with very disappointing results; except for a few cases they found land-improving investments to be an insignificant factor in determining yields. Similar results are also reported by Hayes et al. (1997) for Gambia. Household-specific constraints (credit constraints, labour constraints due to rising migration opportunities) as well as lack of adequate data were often indicated as plausible reasons for not finding significant productivity effects of soil and water conservation.

Here we examine the productivity of indigenous soil and water conservation investments in the Boukombé region in north-west Benin, using an in-depth survey among 101 farmers on farm inputs, outputs and soil and water conservation (SWC) investments. We show that positive effects of SWC investments are only observed if one controls for household-specific constraints. We use a production function approach to relate SWC to farm output, and we control for observable and unobservable household characteristics with household fixed effects. The results show that (1) there are large productivity effects of indigenous SWC investments in the Boukombé region of Benin, and (2) there is positive interaction between plot size and SWC on productivity.

The structure of this chapter is as follows. We start with a discussion of the role of indigenous soil and water conservation techniques for agricultural intensification. It will be argued that there is a strong rationale behind the promotion of indigenous SWC techniques in areas with unfavourable natural conditions and poor infrastructure. In the next section we discuss the different types and adoption rates of the most important indigenous SWC techniques that can be observed in

---

1   This chapter is an abbreviated version of Adégbidi et al. (2004).

the Boukombé region of north-west Benin. In the next section we use a production function approach to relate farm output to inputs and SWC investments to measure the productivity of indigenous SWC investments, before concluding.

## Farmer-based Innovations and Modern Technologies for Agricultural Intensification in LDCs: Key Issues and Perspectives

Many observers have noted that there are few signs of technological progress in many rural areas of LDCs with most of the farmers being still unable to adopt highly mechanized tools, chemical inputs, high-yielding varieties, modern irrigation systems and other, 'industrial', techniques supplied by the green revolution. The extremely low adoption rates of these new methods, techniques and tools, especially in the less-favoured areas in Africa, have renewed the interest of development workers and researchers in farmer-initiated innovations (Spencer 1994; Fan and Hazell 1999). It has been found that several indigenous technologies are already being implemented more or less intensively to combat erosion and to control soil fertility, such as ridging and tie-ridging, silt traps, stone-walled terraces, stone bunds, use of organic fertilizer (green and farmyard manure) and tree planting. Some of these land investments were already in use in the more densely populated parts of sub-Saharan Africa prior to the colonial period (Allan 1965; Miracle 1967; Gleave and White 1969; Morgan 1969; Okigbo 1977; Pingali and Binswanger 1984, 1988; Templeton and Scherr 1997).

Many of these indigenous types of soil and water conservation methods and techniques have been found in Boukombé already in 1959 by the first French soil scientists, Messrs Fauck and Maignien, who were appointed to evaluate the extent of soil degradation, to analyse the available soil conservation techniques and to suggest new methods to arrest erosion and improve soil fertility in the area. They report that stone bunds and ridging were widely used by farmers although they strongly argued that these should not be considered as conservation techniques (translated from French)(Fauck and Maignien 1959, p.5):

> The Sombas hardly apply any indigenous soil conservation methods. The existence [in the area] of stone bunds and pseudo contour bunds might likely delude us. In fact, these [structures] are nothing other than simple heaps of stones collected from the fields. Often, the bunds are constructed following the direction of the highest slope. Some sporadic horizontal bunds are observed but they hardly reduce the water run-off; they often have many breaks, which are filled up by various thin materials from the field. [...]

> The traditional crops are sorghum and hungry rice [which are] grown in every part of the territory where enough space for cultivation is available. All the hills are cultivated and it is likely that nothing is able to speed down the population pressure [on the land resources]. Sorghum is planted on ridges; the more deficient

is the drainage the higher are the ridges [...] The duration of the fallow is very short.

It is worth remembering that the key conceptual lines of the first and most famous integrated agricultural development project implemented in Boukombé in the 1960s (1963–69), the so-called Projet SEDAGRI,[2] were based on the conclusions of this report. And indeed, no attention was given to the indigenous soil conservation techniques by the project, which followed more or less the recommendations of the soil scientists Fauck and Maignien. The project thus put all its efforts into the construction of contour bunds (*banquettes*) made by means of heavily mechanized equipment and hired labour. The second largest activity of the project was the dissemination of the use of chemical fertilizers. In spite of these activities, the region of suffered severe food insufficiency in the late 1970s and early 1980s.

The approach used in the project finds its roots in the Malthusian view that farmers in less-developed countries (LDCs) are unable to show sufficient capacity for innovations when confronted with population pressure, degradation of natural resources and crop yield decline. This line of thought was followed by almost all colonial officers and until very recently by almost all agricultural extension and research administrations in LDCs. Opposite to this view is the Boserupian view of agricultural development, according to which farmers in LDCs are extremely 'dynamic', and the techniques and methods they develop to cope with soil degradation and declining crop yields are appropriate on economic and sustainability grounds (Richards 1939; London and Nadel 1942; Boserup 1965; Reij et al. 1996a; Clay, Reardon and Kangasniemi 1998; Pender 1998). According to this view, local invention and on-farm produced conservation inputs should be sufficient to lead to a significant improvement in land and labour productivity in tropical agriculture. The reasoning is based on the deep-rooted belief that 'the output of land responds far more generously to an additional input of labour than assumed by neo-Malthusian authors' (Boserup 1965, p. 14), and consequently a sufficient increase in farm output could be obtained without applying large quantities of industrial inputs even in cases of very rapid population growth. It is this argument which has given birth in the last two decades to the concept of Low External Input Agriculture (LEIA). This line of thinking was reinforced in the early 1990s following worldwide evidence of soil and environmental damage caused by excessive use of chemical fertilizers and pesticides, irrigation water and tractorization. One can now find many examples in the literature of waterlogging and salinization, fertilization and pesticide contamination of water, increasing pest resistance and resurgence, habitat loss, soil erosion and tapering and even

---

2    SEDAGRI is the name of the French consultancy firm which was in charge of the implementation of the project. The project was implemented three years after Independence in 1960 and at that time the national agricultural extension and research administration had not yet been created.

declining yield potential (Costin and Coombs 1981; Randall 1981; Singh et al. 1987; Subba Rao et al. 1987; Chopra 1989; Pingali et al. 1990; Scherr and Hazell 1994; Fan and Hazell 1999).

Recently, an intermediate line of thought has emerged, arguing that improvements in agricultural productivity and sustainable resource management require complementarity in technology use. Local technical invention should be catalysed and supported and this is particularly necessary in less-favoured areas where natural conditions are unfavourable and infrastructure provision (roads, education, communications, research and extension, etc.) often highly insufficient (Scherr and Hazell 1994). Nevertheless, it is also implicitly or explicitly thought that farmer innovations would have insufficient impact on productivity growth, endangering food security in case of rapid population growth. 'Rapid growth in food output would be achieved only when farmer-initiated innovations are complemented by science- and industry-based inputs, such as high-yielding varieties, fertilizers, pesticides, and similar' (Pingali and Binswanger 1988).

The above discussions suggest that one should study the potential of local technical knowledge and invention (since they are likely to be the most accessible ones) before moving on to more industry-based agricultural practices. By adopting such sequence in policy implementation, one may be able to find a more optimal mix between traditional and new methods, techniques and tools at a more favourable private and social costs/benefits ratio. Bunch and Lopez (1995) report examples of extension programmes in hilly and mountainous areas of Central America which have attempted to accelerate the process of agricultural intensification by strengthening local capacity for innovations (attitudes, institutions, skills in experimentation) rather than focusing on technology introduction (Bunch and Lopez 1995). Unfortunately, tools for assessing the performance of the farmer-initiated soil conservation techniques remain very weakly developed and the empirical evidence is still rare (Boserup 1965; Scherr and Hazell 1994) although it is generally assumed that they improve the productivity of the land.

Later in this chapter we will study the impact of traditional land improving techniques on productivity using plot-level data derived from a questionnaire survey conducted in 1999 in Boukombé (in the north-west of Benin). Regression results indicate that most of these techniques do yield high returns but only when one corrects for household fixed effects. Also we find important differences in productivity across plot size. First, however, we discuss the adoption pattern of indigenous soil and water conservation techniques in the Boukombé region in north-west Benin.

## Adoption of Indigenous Soil and Water Conservation Techniques in Boukombé

Only very recently, with the start of the World Bank project PGRN (Projet de Gestion des Ressources Naturelles) in 1994, have indigenous soil and water conservation techniques been put on the agricultural research and extension agenda in the Boukombé region. The Benin agricultural administration has always been reluctant to initiate research on traditional land improving techniques in this area following the recommendations of Fauck and Maignien (1959).

The PGRN project has conducted a large study on indigenous SWC techniques in Boukombé in 1994 (Kodjo et al. 1995). However, detailed data on the rate and intensity of adoption as well as the spatial distribution of these techniques have not been reported. In order to assess these important aspects and also to evaluate the potential of these techniques in enhancing farm output, we used an in-depth survey among 101 randomly selected farm households in four villages (Takouanta, Okouaro, Kounakogou and Koutagou). Village surveys (PRA) were conducted in July to August 1998 supplemented by a questionnaire survey in January to February 1999. Data on household characteristics, production systems and soil conservation activities were collected for the crop season 1998/1999.

Table 5.1 briefly describes the productive resource endowments and the major sources of income of the households included in the study sample. The vast majority of the households rely mainly on crop production for their subsistence. The market value of the annual gross output from that activity is low (FCFA 111,700 or US$151.31, on average). However, most of the households are also engaged in small-scale livestock farming (guinea fowl, poultry, pig, duck, etc.). About one-third of the households also earn additional income from the sales of gathering products and fruits (firewood, shea nuts, cashew nuts, mangoes, etc.), non-farm employment (pottery, small-scale processing of gathering products and agricultural crops) and remittances. Many households own some goats, sheep or cattle mainly for consumption-smoothing purposes; about 71 per cent own five to six goats or sheep and 41 per cent have four heads of cattle, on average. Until recently, official sources often qualified the economic situation in Boukombé as absolutely desperate; this view was commonly shared in the 1960s and 1970s. Indeed, households in the area face a particularly hostile natural environment; the total area of the sub-prefecture is 1036 sq. km of which only one-third is cultivable for a total population estimated at 69,852 in 2000 (INSAE 2000; Babatoundé and Sounkoua 1996; CARDER 1969–99). Most of the soils are very rocky and increased population pressure has substantially reduced the quality of a large proportion of the cultivable land.

In the 1960s and early 1970s, resettlement programmes were initiated in order to remove part of the population from the most hostile fringes of the region, but most of these initiatives have failed with many people refusing to leave their homeland (Natta 1999). In the second half of the 1990s, official sources indicated an improvement in cereal production growth, renewing the interest of policy-makers in the agricultural potential of the area. Three reasons are usually stressed to explain recent improvements. First, there has been a dramatic increase in the adoption rate of maize, which is being substituted for some traditional crops such

**Table 5.1 Household characteristics, economic activities and income in the sub-prefecture of Boukombé (Northwest Benin)**

| Assets | Takouanta (N=25) | Koutagou (N=25) | Kounakogou (N=26) | Okouaro (N=25) | All villages (N=101) |
|---|---|---|---|---|---|
| Quality of the land at the village levela | very low | low | low to medium | medium to good | - |
| Average farm size (in hectares) | 4.13 | 2.00 | 2.70 | 4.36 | 3.29 |
| Average area cultivated (in hectares) | 1.68 | 1.30 | 2.47 | 3.37 | 2.21 |
| Average number of goats and sheep, mean of the non-zero reported values b | 5 (52%) | 5 (68%) | 6 (88%) | 6 (76%) | 5 (71%) |
| Average number of cattle, mean of the non-zero reported values b | 4 (28%) | 4 (48%) | 4 (42%) | 5 (44%) | 4 (41%) |
| Ownership of ox-plough (% of households) | 0 | 0 | 0 | 28 | 7 |
| Household population | | | | | |
| Average household size | 5 | 7 | 6 | 7 | 6 |
| Active population, 15-59 year (%) | 57 | 52 | 51 | 43 | 54 |
| Average age of the household head (year) | 49 | 47 | 47 | 44 | 46 |
| Gross annual farm output c in fcfa d, sample mean | 69,542 | 97,425 | 106,485 | 173,546 | 111,700 |
| Other sources of income (percentage of households engaged) | | | | | |
| Sales of animal products (chicken, guinea fowl, eggs, honey, sheep, goat, etc.) | 16 | 60 | 50 | 24 | 38 |
| (Distress) sales of cattle | 0 | 8 | 0 | 4 | 3 |
| Sales of gathering products and fruitse | 48 | 4 | 100 | 0 | 39 |
| Non-farm self-employment f | 24 | 28 | 27 | 16 | 24 |
| Temporary migration g | 0 | 4 | 0 | 12 | 4 |
| Wage labour employment | 12 | 4 | 0 | 0 | 4 |
| Migration | | | | | |
| Incidence of migration (% of households with at least one migrant) | 16 | 72 | 16 | 4 | 28 |
| Remittances (fcfa), mean of the non-zero reported values | 22,500 | 46,433 | 46,800 | 0 | 42,520 |
| Connection to roads and markets | bad | medium to good | medium | good | – |

# Notes to Table 5.1

[a] This is compared to the average quality of the land at the level of the whole sub-prefecture of Boukombé.

[b] In the brackets are indicated the percentage of households which own at least one animal.

[c] We include only the main food crops and the commercial crops grown in the study area (sorghum, hungry rice, millet, maize, rice, bean, groundnut, bambara groundnut, yam, cotton, tobacco). The value of the output is imputed by using the sample median of the selling prices, which have been reported by the surveyed households. The percentage of households which have reported some sales of food crops during the year of the study are as follows : sorghum (28% of growers), hungry rice (34%), millet ( 3%), maize (22%), rice (39%), bean (29%), bambara groundnut (12%), yam (9%). The food crops whose production are not included in the total annual gross output are: gumbo, pepper, sweet potatoes, tomatoes and cassava; most of them are cultivated by a negligible number of households (1 per cent); the only exception is gumbo which is grown by 7 per cent of the household but it often occupies a negligible portion of the cultivated land. Note, also, that reliable price figures are not available for these crops in the study area.

[d] US$1 = FCFA 738.23

[e] These include firewood, fruits of wild trees such as shea (*Vitellaria paradoxa*), baobab (*Adansonia digitata*), *Parkia Biglobosa* as well as fruits of planted trees like cashew and mango trees.

[f] These include small-scale processing activities (production of the very popular traditional beer called *tchoukoutou* made of sorghum, production of shea butter, etc.), small-scale trade and handicrafts (mainly pottery). Female household members are the most employed in these activities.

[g] Mainly young household members (aged between 17-25 years) are engaged in these activities. They leave the household as soon as the hardest on-farm works have been completed (often after the weeding activities) to seek a temporary farm employment in the most important cotton growing areas of Benin (e.g. in the provinces of Borgou in the North-east, and Zou in the centre of Benin); sometimes, they travel to the neighboring countries (Ghana, Togo, Nigeria) where they are also employed mainly in the farm sector.

*Source* : Survey UNB/VU (1999)

as hungry rice.[3] Second, there has been a high increase in the rate of application of mineral fertilizers; in 1999, 74,900 kg of fertilizers were purchased in the sub-prefecture while in 1978 only 26,659 kg were used. Third, a large adoption of animal traction has been recorded in recent years in the most fertile areas of the sub-prefecture; the number of ox-ploughs was 167 in 1996 while only three oxen could be found in the area in 1978 (CARDER 1969–99; Adégbidi et al. (2001).

Within the framework of our research, we found that many indigenous SWC techniques are at the heart of the farming systems in the study area, and in all four villages, farmers were found to be well accustomed to the broad patterns of traditional land improving techniques observed in most of the hilly and mountainous areas around the world: structural conservation measures, tie-ridging and ridging, water catchments and biological measures (use of manure, fallowing and crop sequencing). Table 5.2 gives the rate and the intensity of adoption of each of these techniques per village. We also show the level of application of the non-indigenous SWC technologies introduced by the extension and research administration as well as projects[4] financed mainly by international organizations; these include mainly contour bunds, contour ploughing, tree planting, composts, mineral fertilizers,and recently the construction of live barriers (made of *Vetiveria zizanioides*).[5]

Table 5.2 shows that many SWC techniques are used by the farmers in the study area, but also that adoption varies across plots, households and villages. In the next section we will use this variation to identify the impact on productivity of these SWC techniques.

## An Empirical Model for Estimating the Returns from Indigenous Soil and Water Conservation

*Production function approach*

We use a production function approach to analyse the productivity of SWC investments on farm output. We estimate a translog production function, relating farm output to farm inputs and other factors affecting productivity:

$$\ln(output) = \alpha_0 + \sum_i \alpha_i \ln X_i + \sum_i \sum_{j \geq i} \alpha_{ij} \ln X_i \ln X$$

---

3   The area of maize has increased by a factor of 12 between 1969 and 1999.

4   The following projects were implemented to prevent soil and environmental degradation in Boukombé: SEDAGRI [1963–69], UNSO [1984–89], UNDP/FAO [1986–90], PGRN [1995–98], PILSA [1996–98], SNV [1995 up to now], Sasakawa Global 2000 [1992–96].

5   These techniques are discussed in more detail in Adégbidi et al. (2004).

**Table 5.2** **Rate and intensity of adoption of the soil and water conservation techniques in Boukombé**

| | | Takouanta | Okouaro | Kounakogou | Koutagou | All villages |
|---|---|---|---|---|---|---|
| **Indigenous Techniques** | | | | | | |
| Stone bunds | % households | 96 | 0 | 8 | 84 | 47 |
| | % plots | 59 | 0 | 2 | 55 | 23 |
| | % land | 69 | 0 | 2 | 62 | 23 |
| Tie-ridging 1 | % households | 52 | 36 | 92 | 48 | 57 |
| | % plots | 33 | 14 | 72 | 27 | 34 |
| | % land | 24 | 16 | 68 | 19 | 33 |
| Tie-ridging 2 | % households | 4 | 4 | 50 | 4 | 16 |
| | % plots | 1 | 1 | 19 | 2 | 5 |
| | % land | 1 | 1 | 24 | 3 | 8 |
| Water catchments | % households | 28 | 0 | 27 | 44 | 25 |
| | % plots | 11 | 0 | 15 | 18 | 9 |
| | % land | 9 | 0 | 9 | 19 | 7 |
| Animal manure | % households | 16 | 24 | 4 | 12 | 14 |
| | % plots | 5 | 5 | 1 | 5 | 4 |
| | % land | 3 | 4 | 1 | 2 | 3 |
| Green manure | % households | 24 | 20 | 46 | 4 | 24 |
| | % plots | 13 | 6 | 40 | 1 | 14 |
| | % land | 14 | 10 | 42 | 0 | 19 |
| Mulching | % households | 0 | 12 | 0 | 0 | 3 |
| | % plots | 0 | 3 | 0 | 0 | 1 |
| | % land | 0 | 5 | 0 | 0 | 2 |
| Fallowing | % households | 80 | 32 | 19 | 64 | 48 |
| | % plots | 38 | 5 | 5 | 20 | 17 |
| | % land | 58 | 19 | 9 | 35 | 31 |
| Crop-rotation | % households | 32 | 38 | 62 | 80 | 60 |
| (cultivation of | % plots | 9 | 11 | 16 | 26 | 15 |
| hungry rice) | % land | 10 | 10 | 20 | 40 | 17 |
| **Non-Indigenous Techniques** | | | | | | |
| Tree planting | % household | | | | | |
| | No planted trees | 48 | 60 | 54 | 52 | 54 |
| | 1-5 trees | 16 | 16 | 35 | 40 | 27 |
| | > 5 trees | 36 | 24 | 12 | 8 | 20 |
| Live barrier, vetiver | % households | 0 | 0 | 4 | 0 | 1 |
| | % plots | 0 | 0 | 2 | 0 | 1 |
| | % land | 0 | 0 | 2 | 0 | 1 |
| Fertilizers | % households | 32 | 36 | 27 | 44 | 35 |
| | % plots | 10 | 11 | 10 | 18 | 12 |
| | % land | 8 | 11 | 12 | 22 | 12 |
| Compost | % households | 4 | 20 | 12 | 0 | 9 |
| | % plots | 1 | 3 | 9 | 0 | 4 |
| | % land | 1 | 3 | 7 | 0 | 3 |
| Contour bunds | % households | 0 | 0 | 23 | 2 | 8 |
| (banquettes) | % plots | 0 | 0 | 8 | 4 | 3 |
| | % land | 0 | 0 | 8 | 3 | 3 |
| Contour ploughing | % households | 0 | 0 | 4 | 4 | 2 |
| | % plots | 0 | 0 | 1 | 3 | 1 |
| | % land | 0 | 0 | 1 | 2 | 1 |

where $\alpha_o$, $\alpha_p$ $\alpha_{ij}$ $\beta_j$ are the coefficients, $X_i$ are the inputs and $Z_i$ are other factors affecting productivity. The translog production function is a flexible specification as it can be viewed as a second-order approximation to an unknown production function (Christensen et al. 1971).

*Definition and summary statistics of regression variables*

Table 5.3 gives the summary statistics for the model variables. **Output** can be measured in physical units (kg) or monetary units (Fcfa). Because we are interested in the (physical) productivity effect of SWC, output is here measured in kilogrammes. Prices may vary widely across years and seasons and the monetary value of output may therefore not always reflect the productivity of SWC. Only plots with cereals (sorghum, millet, maize, hungry rice) and cereals mixed with beans were included to avoid aggregation bias.[6] Because output has been measured in monetary terms in several earlier studies, we will check whether the results are sensitive to the choice of output measure. For the plots in the sample, the average output is about 200 kilogrammes. Given that the average plot size is approximately 0.60 hectares, the average output per hectare is 344 kg.[7]

**Crop shares** for sorghum, millet, maize and beans indicate the presence of these crops as a proportion of the total number of crops grown on a plot.[8] These shares will be included in the regression analysis to control for output heterogeneity.

With respect to the inputs, **labour** is measured as the total amount of full labour days used for clearance, tillage, planting, weeding, harvesting, applying fertilizers and/or pesticides, and manuring and composting. Family labour, hired labour and exchange labour are included. Labour is standardized into adult equivalents, where 1 is for adult men, 0.75 for adult women, and 0.25 for children. **Land** is given in hectares, and the plots are on average 0.60 hectares. The average plot receives about 54 full days of equivalent labour per cropping season. The amount of **fertilizer** is given in kilogrammes. On average farmers use approximately 10 kg of fertilizer.[9] We have also included the average number of **livestock** per plot

---

6    On average, more than 2/3 of the total cultivated land are planted with sorghum, millet, hungry rice, maize and beans per household.

7    This is much lower than the average yields reported by CARDER, which is responsible for gathering data on farm output in Benin. Similar low yields have been found for the Atacora (surveys UNB/VU 1995–96, research project on 'Soil Degradation in Benin: Farmers' Perceptions and Rational Behaviour', sponsored by CREED) and for southern Benin (surveys UNIHO-G5 1994–98, research programme on 'Participatory Technology Development for Soil Fertility Restoration in Southern Benin', sponsored by GTZ). Unfortunately none of the CARDER reports explains the derivation of their published yields (Gandounou and Adégbidi 2000).

8    So for instance a crop share of 0.5 for sorghum indicates that sorghum is one of the two crops cultivated on a plot. The crop shares reduce to crop dummies if there is no multi or mixed cropping.

9    Fertilizer is applied to 15 per cent of the plots included in the regression analysis.

because the manure functions as an input in the production system. The amount of livestock may also be a proxy for wealth. Wealth may affect production given that credit markets are imperfect and wealthy farmers may have better access to credit and less hesitant to use it. The total livestock is measured in Tropical Livestock Units (cow and oxen = 1.0, sheep and goats = 0.20), and amounts to 0.53 livestock unit per plot. The dummy variable **manure** equals one if farmers use manure and zero otherwise, and 8 per cent of the plots receive manure. This discrete variable is sometimes used instead of the continuous livestock variable, because it varies by plot within the household.[10] On 8 per cent of the plots farmers use **animal traction**, and it may be assumed that this affects production positively (although it may also increase erosion).

In terms of plot characteristics, the dummy variable **high plot fertility ranking** indicates whether the farmer ranks the fertility level of his plot among the top half of his plots. It may be assumed that plots of higher fertility are more productive. The dummy variables **rocky soil** and **gravelly soil** indicate whether the plot is on rocky, respectively gravelly soil. Sixty six per cent of the plots have either rocky or gravelly soil, which should be less productive than the omitted category of sandy, loamy, and/or clay soil. The dummy variables **light slope** and **steep slope** indicate whether the plot is respectively on a slope up till 10 per cent or on a slope in excess of 10 per cent.

The steeper the slope, the more problems there will be with soil erosion and runoff, and the less productive the plot will be. In the sample 36 per cent of the plots are on light slopes, and 17 per cent of the plots on steep slopes. The **distance to home** variable is measured in metres, and is 560 metres on average. Because farmers will try to avoid travel time and transportation costs, plots which are located further from the *tata* or homestead may receive less attention . The last plot characteristic we have included is the dummy variable **plot is borrowed/rented**. The effect of tenure on productivity is potentially negative – farmers may be less motivated to improve borrowed or rented plots. On average 14 per cent of the plots are borrowed or rented.

With respect to indigenous soil and water conservation techniques we have included four dummy variables. The first is **stone bunds** and Table 5.3 shows that 28 per cent of the plots have these soil conservation structures. The second soil and water conservation dummy variable is for the presence of **tie-ridging (variant 1)** on the plot. This is the physical conservation technique which has been most frequently adopted (41 per cent). The third soil and water conservation dummy variable is for the **other type of ridging (tie-ridging variant 2, contour ploughing or contour bund)** which has been adopted on 8 per cent of the plots. Here we have combined indigenous and non-indigenous techniques because they are very similar and each of them is used on very few plots. The last variable is for the presence of a **water catchment** on the plot -11 per cent of the plots have this

---

10   This is important if we want to control for household fixed effects.

**Table 5.3     Descriptive statistics of model variables**

| Variable | Obs | Mean | Std. Dev. | Min | Max |
|---|---|---|---|---|---|
| **Output** | | | | | |
| output (kg) | 260 | 206.64 | 161.07 | 10.00 | 1100.00 |
| sorghum (share) | 260 | 0.36 | 0.40 | 0.00 | 1.00 |
| millet (share) | 260 | 0.18 | 0.33 | 0.00 | 1.00 |
| maize (share) | 260 | 0.13 | 0.32 | 0.00 | 1.00 |
| beans (share) | 260 | 0.10 | 0.19 | 0.00 | 0.50 |
| **Inputs** | | | | | |
| labour (equivalent labor) | 260 | 54.49 | 30.72 | 4.50 | 205.94 |
| land (ha) | 260 | 0.60 | 0.49 | 0.13 | 4.00 |
| amount of fertilizer (kg) | 260 | 9.88 | 26.36 | 0.00 | 150.00 |
| number of livestock (TLU) | 260 | 0.53 | 0.65 | 0.00 | 5.33 |
| use of manure | 260 | 0.08 | 0.28 | 0.00 | 1.00 |
| use of animal traction (dummy) | 260 | 0.08 | 0.28 | 0.00 | 1.00 |
| **Plot characteristics** | | | | | |
| high plot fertility ranking (dummy) | 260 | 0.46 | 0.50 | 0.00 | 1.00 |
| rocky soil (dummy) | 260 | 0.22 | 0.41 | 0.00 | 1.00 |
| gravelly soil (dummy) | 260 | 0.44 | 0.50 | 0.00 | 1.00 |
| light slope (dummy) | 260 | 0.36 | 0.48 | 0.00 | 1.00 |
| steep slope (dummy) | 260 | 0.17 | 0.38 | 0.00 | 1.00 |
| distance to home (km) | 260 | 0.56 | 0.88 | 0.00 | 5.00 |
| plot is borrowed/rented (dummy) | 260 | 0.14 | 0.35 | 0.00 | 1.00 |
| **Soil and water conservation techniques** | | | | | |
| Stone bunds (dummy) | 260 | 0.28 | 0.45 | 0.00 | 1.00 |
| Tie-ridging (dummy) | 260 | 0.41 | 0.49 | 0.00 | 1.00 |
| Other type of ridging (dummy) | 260 | 0.08 | 0.28 | 0.00 | 1.00 |
| Water catchment (dummy) | 260 | 0.11 | 0.32 | 0.00 | 1.00 |
| **Household characteristics** | | | | | |
| number of plots | 260 | 6.08 | 3.10 | 1.00 | 17.00 |
| Village fixed effects | | | | | |
| Takouanta (dummy) | 260 | 0.24 | 0.43 | 0.00 | 1.00 |
| Okouaro (dummy) | 260 | 0.30 | 0.46 | 0.00 | 1.00 |
| Kounakogou (dummy) | 260 | 0.30 | 0.46 | 0.00 | 1.00 |

type of physical structure. It is assumed that each of these techniques has a positive impact on productivity.

For household characteristics we include the **number of plots**. If farms are very fragmented, farmers will be less able to take advantage of possible economies of scale at the plot level (such as ploughing, manuring). We also include in the regressions village dummies to control for other, not included, village differences, such as differences in overall soil fertility, differences in knowledge and farming tradition, the access to exchange labour and distance to markets.

## Empirical Results

*Bivariate analysis of yields and use of soil conservation techniques*

First we present a number of cross tables which give the mean plot yield by adoption of soil conservation technique for each of the villages. Next we will apply multivariate regression techniques to estimate an agricultural production function for plot output taking into account inputs, plot characteristics, adoption of soil and water conservation techniques, household and village characteristics. Plot output is measured as the total number of kg per plot, and plot yields are measured as plot output per ha. We have only included plots on which the following cereals were grown: sorghum, millet, maize, hungry rice. We have also included plots for which these cereals were grown as well as beans because many plots have a mixture of millet or sorghum with (some) beans.

Table 5.4 reports the difference in yields between plots with and without indigenous soil and water conservation techniques. On average for all villages (last column), we see that plots with tie-ridging, variant 1 (*billonnage cloisonné*), water catchment, fertilizer and manure do better than plots without these conservation structures. No clear effect is found for stone bunds, and actually plots with other type of ridging do *worse* than plots without these conservation structures. There are differences between the villages, however, such that, in some villages, plots with a technique are more productive and in others less productive than plots without the technique.

The most likely explanation for this ambiguity is that there are differences in plot characteristics between plots with and without the soil and water conservation technique. In particular, plots that suffer from erosion, lack of water retention or low fertility in the first place may be more likely to receive soil and water conservation investments than plots which have no such problems. In these instances farmers will be more motivated to try to improve these plots. Still, the table suggests that soil and water conservation techniques are productive as we observe higher productivity for most of the indigenous soil and water conservation techniques.

**Table 5.4**     Adoption of indigenous techniques and average plot yield (kg/ha). Number of plots within brackets

| Technique | Takouanta | Okouaro | Kounakogou | Koutagou | All villages |
|---|---|---|---|---|---|
| **stonebunds** | | | | | |
| yes | **310** (43) | | **278** (3) | **688** (28) | **452** (74) |
| no | **525** (19) | **416** (81) | **452** (74) | **562** (17) | **454** (191) |
| **tie-ridging, variant 1** | | | | | |
| yes | **513** (22) | **306** (12) | **453** (60) | **718** (13) | **481** (107) |
| no | **301** (40) | **436** (69) | **416** (17) | **609** (32) | **435** (158) |
| **other type of ridging** | | | | | |
| yes | | | **440** (16) | **422** (6) | **435** (22) |
| no | **376** (62) | **416** (81) | **446** (61) | **674** (39) | **455** (243) |
| **water catchment** | | | | | |
| yes | **371** (12) | | **700** (2) | **919** (15) | **677** (29) |
| no | **378** (50) | **416** (81) | **438** (75) | **501** (30) | **426** (236) |
| **fertilizer** | | | | | |
| yes | **638** (8) | **639** (18) | **780** (5) | **606** (10) | **648** (41) |
| no | **337** (54) | **353** (63) | **422** (72) | **650** (35) | **418** (224) |
| **manure** | | | | | |
| yes | **544** (5) | **424** (9) | **500** (7) | **1500** (2) | **567** (23) |
| no | **361** (57) | **415** (72) | **439** (70) | **601** (43) | **443** (242) |

*Multivariate regression analysis*

The above bivariate tables suggest that there are productivity effects from soil conservation in the region of Boukombé in the north-west of Benin. Plots which have tie-ridging, water catchments, fertilizer or manure show higher yields on average. Plots with stone bunds and other types of ridging, however, do not show higher yields than plots without these measures. The interesting question becomes therefore whether we can find productivity effects if we would control for other, possibly intervening factors. Plots that are steeper might receive more soil conservation investments, although they are less productive than plots with less slope. Or farmers might choose to invest in fields which lack organic matter or minerals.

We estimate the translog production function both with OLS and with household fixed effects. The productivity of farm production may be affected by household-level factors, such as farming skills and labour and credit constraints, and given that we have multiple plot-level observations for each household we can control for this by including household dummies.

The first two columns of Table 5.5 report the regression results if we use ordinary least squares (OLS) to estimate the translog production function, while the last two columns report the fixed effects results. We do not include all interaction terms because only the interaction terms for land with the soil and water conservation investments were found to be significant at 5 per cent in the fixed effects regression.[11] This suggests that soil and water conservation affects the productivity of land rather than the productivity of labour. This result confirms what has been found by Byiringiro and Reardon in their study of farm productivity in Rwanda (Byiringiro and Reardon 1996). In particular they found that farms with higher investment in soil conservation have much better land productivity than average, but no significant effect was found for labour productivity.

Significant positive effects on physical cereal output are produced by labour, land, the amount of fertilizer, number of livestock and the use of animal traction in the OLS regression. Similarly, there is a significant negative effect on production if the plot is on a steep slope and plots on light slopes appear also to be less productive.

Importantly, however, the OLS results show no significant effects from SWC except in terms of water catchments. The fixed effects regression (last two columns of Table 5.5) shows that the effect of soil conservation on productivity becomes very clear if we also control for the unobserved household characteristics. The interaction terms of land with stone bunds, tie-ridging and other type of ridging are all positive and significant suggesting that productivity from stone bunds, tie-ridging and other ridging does increase with plot size.[12] The same results are found if the dependent variable of the regression is the value of all crops. Larger plots benefit more from SWC because they tend to be less protected by barriers (such as trees and ditches) and therefore more prone to erosion. The estimates suggest that the total productivity effect of some of the SWC investments may actually have been negative for the smaller plots in the survey year.[13] This may reflect the fact that smaller plots are relatively better protected against erosion and that SWC measures tend to take up a relatively larger proportion of otherwise productive land on smaller plots. Interestingly the translog production function estimates obtained by Byiringiro and Reardon for Rwanda show the same pattern with the total productivity effect of SWC as possibly negative for small farms.[14]

---

11   They were also insignificant in the OLS regression.

12   The interaction term between land and water catchments was insignificant and has been omitted from the regression.

13   Note that the term ln(land) becomes negative for small plots.

14   Byiringiro and Reardon (1996), Table 2. Note that according to the reported estimates by Byiringiro and Reardon total productivity may be an increasing function of erosion for small plots (ln(land) is negative). Because the 'erosion' variable in the estimated translog production function is an inverse function of soil conservation practices an increase in soil conservation may therefore actually lower total productivity. Byiringiro and Reardon do not note this point because they look at *marginal* productivities.

**Table 5.5    Production function estimates. Dependent variable is in(output). OLS and household fixed effects regressions including interaction terms with SWC investments**

| Variables | OLS | | Fixed Effects | |
|---|---|---|---|---|
| | *coefficient* | *t-value* | *coefficient* | *t-value* |
| **Variable inputs and medium-term land investments** | | | | |
| ln(labour) (equivalent labour) | **0.18** | 1.90 | **0.25** | 2.28 |
| ln(land) (ha) | **0.27** | 2.42 | -0.08 | -0.56 |
| amount of fertilizer (kg) | **0.004** | 2.35 | **0.004** | 2.05 |
| number of livestock (TLU) | **0.16** | 2.87 | | |
| use of manure (dummy) | | | -0.18 | -1.06 |
| use of animal traction (dummy) | **0.35** | 1.67 | 0.08 | 0.41 |
| **Plot characteristics** | | | | |
| high plot fertility ranking (dummy) | 0.08 | 0.80 | 0.12 | 1.38 |
| rocky soil (dummy) | 0.11 | 0.70 | -0.12 | -0.59 |
| gravelly soil (dummy) | 0.07 | 0.55 | -0.18 | -1.23 |
| light slope (dummy) | -0.12 | -1.08 | 0.06 | 0.46 |
| steep slope (dummy) | **-0.24** | -1.67 | -0.13 | -0.74 |
| distance to home (km) | -0.01 | -0.17 | -0.03 | -0.61 |
| plot is borrowed/rented (dummy) | 0.07 | 0.55 | 0.09 | 0.45 |
| **Soil and water conservation techniques** | | | | |
| Stone bunds (dummy) | -0.15 | -0.71 | 0.21 | 0.88 |
| * ln(land) (ha) | 0.20 | 1.44 | **0.73** | 3.80 |
| Tie-ridging (dummy) | -0.15 | -0.88 | 0.01 | 0.05 |
| * ln(land) (ha) | 0.04 | 0.32 | **0.34** | 2.16 |
| Other type of ridging (dummy) | -0.07 | -0.33 | 0.26 | 0.91 |
| * ln(land) (ha) | 0.17 | 0.84 | **0.55** | 2.26 |
| Water catchment (dummy) | **0.35** | 2.28 | **0.41** | 2.35 |
| **Household characteristics** | | | | |
| number of plots | 0.02 | 1.03 | | |
| **Village fixed effects** | | | | |
| Takouanta (dummy) | **-0.48** | -3.25 | | |
| Okouaro (dummy) | **-0.65** | -3.30 | | |
| Kounakogou (dummy) | **-0.22** | -1.43 | | |
| **Household fixed effects** | no | | yes | |
| **Constant** | **4.79** | 10.71 | **4.20** | 8.73 |
| Number of observations | 260 | | 260 | |
| R² | 0.40 | | 0.24 | |
| Significance level regression | 0.0000 | | 0.0000 | |

Standard errors are robust. Coefficients with p-value of 0.10 or less are printed in bold. Crop shares are also included in the regression but not reported.

**Table 5.6    Productivity effect of SWC by plot size (%)**

|  | mean | 90% confidence interval | |
|---|---|---|---|
| Stone bunds | | | |
| 0.5 ha | -0.24 | -0.45 | 0.03 |
| 1.0 ha | 0.26 | -0.16 | 0.76 |
| 1.5 ha | 0.71 | 0.03 | 1.57 |
| Tie-ridging | | | |
| 0.5 ha | -0.19 | -0.38 | 0.06 |
| 1.0 ha | 0.03 | -0.26 | 0.36 |
| 1.5 ha | 0.18 | -0.21 | 0.67 |
| Other ridging | | | |
| 0.5 ha | -0.09 | -0.38 | 0.31 |
| 1.0 ha | 0.34 | -0.17 | 0.98 |
| 1.5 ha | 0.70 | -0.08 | 1.75 |
| Water catchment | | | |
|  | 0.52 | 0.14 | 0.95 |

*Note*: the productivity effect is calculated by comparing the productivity of plots with and without SWC keeping all other variables constant. The means and 90% confidence intervals were estimated from a random sample drawn from the distributions of the estimated regression coefficients (sample size = 1000).

In Table 5.6 we report the estimated productivity effect by different plot sizes. The table shows that for plots of 1.5 ha (above average) the productivity effect of SWC tends to be positive and quite large. The average productivity effects for stone bunds, other ridging and water catchments are 71, 70 and 52 per cent respectively. The estimated effect for tie-ridging is lower at 18 per cent. The confidence interval is quite large, however, with an upper bound of 67 per cent for tie-ridging. The productivity effect for plots of 0.5 ha (below average) is lower and possibly even negative for stone bunds, tie-ridging and other ridging. We note however that confidence intervals tend to be large.

In summary, we find large observable productivity effects of indigenous SWC investments in the Boukombé region of Benin, but only for the larger plots. This raises the interesting question why some farmers have nevertheless invested in smaller plots. First, many of these investments (especially stone bunds) were done in the past when these investments may have been more productive; for instance, because of the absence of alternatives such as chemical fertilizer and drought-resistent crop varieties. Second, farmers indicated in the survey that soil conservation tends to increase the stability of yields which is crucial for farmers close to the subsistence level. Third, the observed productivity effects may be atypical, in the sense that the survey period was atypical. Rainfall during the 1998–99 season was 1470 mm which is 36 per cent higher than the average annual rainfall in the last 76 years. During the planting period (May–June) the total rainfall was also very high (62 per cent higher than the average). Similar figures are also observed for the rainfall intensity (113 mm/day versus an average of 83 mm/day for the whole year; 21 mm/day versus an average of 13 mm/day during the planting period). If productivity effects depend on rainfall, they may have been atypical as well. However, it is difficult to assess whether the observed productivity effects are an over- or underestimate of the average productivity effects. Although many farmers in our study area considered heavy rains as having negative effects on productivity in spite of the very wide application of various mechanical and non-mechanical conservation structures, it may well be possible that without those structures they would have been even worse off. Longitudinal data on soil and water conservation investments in combination with production and rainfall data should shed more light on this issue.

We also tested whether other kinds of interaction terms should be included besides those between land and the indigenous techniques. We have pursued three possibilities, namely that (1) the productivity of stone bunds is higher if they are perpendicular to the slope or if they are reported to be of good quality (as opposed to poor quality), (2) the productivity of soil and water conservation depends on fertilizer use, and (3) there are village differences in the productivity of soil and water conservation.

We do not find strong evidence that perpendicular or good quality stone bunds are more productive although good quality stone bunds may be somewhat more productive (the p-values are 0.06 and 0.77 respectively). Possibly the productivity differences are too small here to be picked up by our relatively small sample. Also the effect of the indigenous techniques does not depend on the use of fertilizer as the p-value for interaction terms between fertilizer use and stone bunds and (other type of) tie-riding is 0.61. Finally we also do not find that some techniques are more productive in some villages relative to others. The p-values of interaction terms between village dummies and stone bunds, tie-ridging, other type of ridging and water catchment are respectively 0.45, 0.17, 0.14 and 0.36.

Finally, we also tested for endogeneity, omitted variable and sample selectivity bias. For instance, one may argue that the coefficients for the soil and water conservation variables may be biased if soil and water conservation practice is endogenous. Also one may argue that our results suffer from an omitted variable bias because we do not control for unobserved plot characteristics. And we have estimated the productivity of indigenous techniques only for plots on which cereals are grown, and therefore we may have introduced a sample selectivity bias. In Adégbidi et al. (2004) we report a number of tests and robustness checks suggesting that these potential biases do not drive the above results.

**Conclusions and Discussion**

In this chapter we have examined the productivity of soil and water conservation techniques as adopted in Boukombé. Farmers adopt a wide range of indigenous

SWC techniques in this area but little has been known so far about the productivity of these techniques. Based on an in-depth survey among 101 farmers, the analysis showed that SWC techniques are indeed found to be strongly productive, but only if one controls for household-specific factors. These household-specific factors may reflect unobserved inputs (such as soil fertility and farmer's knowledge and skill) as well as household-specific constraints on efficient farm production (such as labour and credit constraints). Also there is a positive interaction between SWC and plot size in terms of productivity, implying that the productivity of SWC techniques is larger on larger plots.

Chapter 6

# More Market, Less Poverty, Less Erosion?
# The Case of Benin

Esaïe Gandonou and Kees Burger

## Introduction

The previous chapter provided evidence that the returns to soil and water conservation (SWC) are positive, once one adjusts for farm specific conditions. Among the many specificities of farm households, especially in developing countries, is their extent of involvement in the market. With farms widely scattered, as is typical for Africa, and perhaps even more for Boukombé, and with underdeveloped road infrastructure, many farm households are first of all focused on securing their own subsistence. In doing so, they set their own – virtual – prices for food, labour and thereby for SWC. They do so in conditions that are highly variable, dependent as farm households are on weather conditions. If, as Gandonou and Oostendorp showed in the previous chapter, the SWC investments can only be understood well at the household level, this suggests that household-level choices set the conditions under which these investments are profitable. In this chapter, we investigate important determinants of these household specific conditions, namely the distance to the market and the household size. We show how household that are less isolated, i.e. closer to Boukombé, have shadow prices closer to the market prices, and hence are less dependent on household specific conditions. In the absence of a market, larger households have larger supply of labour to the farm (lower shadow prices of labour) and larger demand for food (higher shadow prices) from the farm. Both factors combine to push labour use on the farm to low levels of marginal physical products. We use simulation to work out what this means for fertilizer use and the attractiveness of soil and water conservation.

In the report submitted by the French soils scientists R. Fauck and R. Maignien to the central authorities of Benin in 1959, they considered Boukombé 'the seat of large scale erosion' with soils being 'overexploited'. This sounds quite similar to the verdict on Machakos that Tiffen et al. (1994) quote. While Machakos has grown into an example of successful recovery, Boukombé and its surroundings have not changed much. In this chapter we shall give an overview of the evolutions in the region and then embark on an analysis of how improved infrastructure helps farmers cultivate more rewarding crops, and – perhaps – farm more sustainably. In the first section we introduce some spatial elements of the region of study and present some demographics. In the next section, we describe the model and the data collection; results are presented in the third section, and in the fourth section we give some estimates of what the effects would be of changes in distance to the market and in population density. Conclusions are in the final section.


### Boukombé

Boukombé is located in the north-west of Benin at a distance of 582 km from Cotonou, the major city of Benin. It is situated at 43 km from Natitingou, the headquarters of the department of Atacora.[1] From Cotonou it takes between eleven and twelve hours by car to reach the headquarters of the sub-prefecture of Boukombé.

Boukombé is populated almost exclusively by the Otamari ethnic group. The Otamari represent also one of the major ethnic groups in Atacora. According to the 2002 census the Otamari form 37.4 per cent of the population of Atacora and 6.1 per cent of the total population of Benin (INSAE 2002).[2] A particular trait of this ethnic group is that, contrary to many other groups in Benin, the Otamari have always had a disinclination to be led by a strong and permanent leader or group of persons. Otamari people are much attached to egalitarianism. In contrast to many places in Benin the households in Boukombé are basically of a nuclear type rather than of the extended family type. Many believe that the Otamari place more value on the individual rather than on social groups. Indeed the adults from the Otamari (those aged more than 18–20 years) strive to be economically independent from each other. According to popular wisdom in these communities it is the responsibility of each individual to secure his own basic economic needs rather than expecting assistance from others. This rather widely admitted social rule does not hold in most ethnic groups in Benin in particular in those areas where the social organization of the community is constructed around extended families and a strong social hierarchy. Many claim that because of their rather strong attachment to individual achievements and weak leadership the Otamari are too individualistic, and this is undermining collective action and development of basic public investments in Boukombé (further discussion on this issue can be found in Natta 1999).

---

1   Boukombé is included into the department of Atacora. In 1999 the administrative division of the territory in Benin was reformed. The country has today twelve departments rather than six. The territory of the department of Atacora is lower now than before the reform. But in this book at any place where we refer to Atacora, note that we mean Atacora 'before the reform'.

2   In other papers this ethnic group is designated by the name Somba. However, recently Benin historians have ascertained that the correct name of this group is Otamari. Meanwhile the name Somba is strongly contested by the members of the ethnic group (Natta 1999).

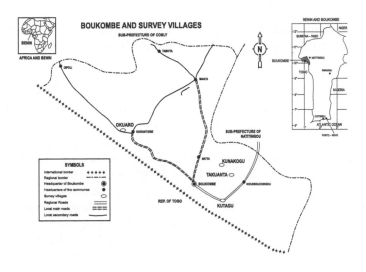

**Figure 6.1    Boukombé and the survey villages**

As in any hilly region, settlement in Boukombé is highly dispersed due to the rugged topography and the implied transport problems. The population of the villages of Boukombé live in dispersed hamlets. Within each hamlet the distances between homesteads (also called *tata)* vary between about 200 m and 1 km. The most striking feature of Boukombé, however, is that even in the plains and less-hilly areas of the sub-prefecture, the households still prefer to install their homesteads at some distance from each other. Indeed, this is in sharp contrast to what is seen in most rural areas in Benin. It seems that there is a strong aversion within the Otamari communities to living in agglomerated settlements. One of the explanations of the relatively high dispersal of the settlement in the area is that households often prefer to have larger home plots. Indeed a popular land occupation practice in Boukombé is to always have home plots of rather a substantial size around each homestead. These plots are seen as more valuable by the farmers. They are subject to intensive animal manuring. Nevertheless there are persistent complaints by development workers who argue that the quasi-absence of agglomerated settlements in Boukombé constitutes one of the key explanations for the low performance of public infrastructure projects (education, health, water supply, roads) in the region.

*Population density and growth*

In 1961, population density in Boukombé was estimated at 35 persons per square kilometre. Population density in Atacora and Benin was estimated at 10 and 18 persons per square kilometre, respectively. Population density in Boukombé was found to be higher than the densities in the remaining sub-prefectures of the northern Benin (INSEE 1961). Since 1961, population growth in Boukombé has been lower than the national average. Using the census data for 1961, 1979, 1992 and 2002, annual growth rates for the three periods in between the censuses can be derived. For Boukombé these amount to 0.4 per cent, 1.6 per cent and 0.4 per cent, whereas for the country as a whole, growth rates were 2.6 per cent, 3 per cent and 3.2 per cent. While differences in fertility can be the reason for this, the more likely explanation is a high degree of out-migration from Boukombé. For the Atacora as a whole, this pattern is not so clear: population growth rates here are about the same as the national average, so that out-migration seems stronger for Boukombé than for Atacora as a whole. By 2002, population densities in Boukombé and Atacora are around 58 and 29 persons per $km^2$.

*Land and soil fertility*

Since 1985, the total area of cultivated land in Boukombé has been rather stable with around 15,000 ha allocated to the main crops. Of these, cereals are by far the most important with 72 per cent of total area. Major grains are sorghum and millet, followed by hungry rice and, more recently, maize. Vegetables and yam form the other important crops. Land quality was assessed to be rather low in the 1950s, and later soil research by Azontondé et al. (1995) confirmed the low quality of soils in Boukombé. Although comparability of the estimates of soil quality over time is low, the two estimates of 1995 and 1960 for the same soil type (Alfisols) indicated a fall in pH and almost a halving of the organic matter. Yet, yields show no declining trend. Figure 6.2 shows the trends in yields of the major crops over the last ten years. There is no clear explanation for this pattern. Rainfall was highly variable from one year to the next, and fertilizer is little used, but use is increasing over time (from 8 kg/ha in 1996 to 16 kg/ha in 2001).

In a survey of 1995, Mulder (2000) reports on the perceptions of farmers as to the quality of the land. In general (97 per cent), farmers felt that soil fertility has declined. The main reasons for this degradation are considered to be deforestation (59 per cent of the households), over-exploitation and bush fires (55 per cent each). Erosion is mentioned by 30 per cent of the households. When asked about what measures these farmers take to improve the quality of their land, the answers differ according to accessibility of new land. Where new land is still easily accessible, expansion of the farm is the favourite response; where access is 'almost impossible' (as typically the case in Boukombé), investment in land quality is the most popular answer (Mulder 2000, p. 73).

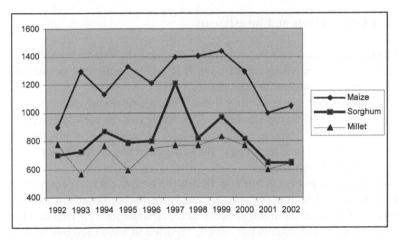

**Figure 6.2     Yields (kg/ha) of major crops in Boukombé**

In the sample used for the present study, these investments take the shape of stone bunds, used by 47 per cent of the households, or stone ridges in more or less connected frames (tie-ridging), used by 57 per cent of the households. Where stones are abundant, the major contribution of these ridges is the clearing of the land of stones, with reducing the erosion and improving the water uptake coming as secondary effects.

*Access to markets*

Agricultural production is mainly subsistence oriented. Mulder (2000) reports on a large sample for Atacora taken in 1995, that only 5 per cent of output of major staples is sold, and only about 3 per cent of the cereal consumption is bought in the marketplace. Of other crops, such as groundnuts, larger shares are sold (28 per cent), whereas rice features prominently among the items bought. Most cash income is earned from sales of farm products, including cattle, but an important part comes from seasonal migration and sales of services and goods other than farm products. The cash income is spent almost completely (97 per cent in the present sample) on non-staples.

The participation in the market for inputs is limited to special forms of exchange of labour and the participation in labour groups that, in exchange for cash and/or food, help prepare land, or weed or harvest the crops. There is no land market as such, though security of tenure is high. Little draught power is used. Physical access to markets is via footpaths or non-tarmac roads. Use of these roads is often difficult during the rainy season.

### A Model for Markets and Investments

The model developed here may represent a typical small farm household from a less-favoured area in a poor country, yet, it is constructed based on our empirical evidence from the Boukombé region.

We assume that the household owns a fixed amount of cultivable land. Its current output includes only one staple. This staple may be viewed as representing several annual food crops as well as products from animal production such as poultry, guinea fowl (and their eggs), duck, pigs and honey. Additional sources of income are very limited.

A local labour market is missing. Hence the household must count in general on its own family labour in farm production. Nevertheless households are still able to obtain off-farm labour through non-market arrangements: invitation or mutual aid. Almost all households in our sample reported to have used one of these two forms of labour contracts. But still the bulk of the labour used consists of family labour.

A land market is also missing. In the sample no sales, purchases or hiring of land were reported. Yet property rights are virtually privatized in the area. Most plots have a well-known proprietor; they are usually inherited. Very few plots are 'newly cleared'. Communal land use rights have mostly been given up in the area some decades ago.

There is a market centre (town) where the household can sell its produce. It can also purchase from the same market food and non-food commodities as well as fertilizer. Food commodities include a staple (which is assumed to be the same as the one produced by the household), oil, the local beer *shukutu*, fish, sugar, etc. Non-food items are mainly soap, kerosene, batteries (for radios), clothes, etc.

The market centre is assumed to be supplied with all kinds of commodities (food and non-food) from the national and/or world markets. 'Urban' households and traders (wholesalers, assemblers, etc.) are the major consumers in that market.

All the farm households face the same market prices but each of them has to pay specific transactions costs. In this framework, we have a broad view of transaction costs. They are supposed to include the easy-to-observe transportation costs but also information, screening and negotiation costs (as also done by Mwakubo et al. in this volume). We further simplify the analysis by postulating that the staple (as defined above) is the only commodity for which transaction costs are important. Indeed in the context of the study area, most products defined as staple are bulky commodities for which the value/weight ratio is in general low compared to other goods (e.g. non-staple consumer goods, fertilizer, etc.) (see also Minot 1999).

In this framework transaction costs are assumed to differ substantially between households even within the same village. The land occupation system in Boukombé region is similar to the one observed in most hilly and mountainous areas around the world. The homesteads are often highly scattered because steep slopes dominate the landscape, high dispersion of the cultivable land and transportation problems.

All expenditures on the market-purchased commodities must be paid in cash. Cash income may come from the non-farm sources as mentioned above (e.g. remittances, income from non-farm activities) or from the sales of the farm output. The majority of households rely, however, on the cash income they can earn from the sales of their output in order to purchase fertilizer.

The main problem the representative farm household is facing is that its yields stagnate or improvements are limited. As a result output per capita stagnates or even decreases. To change this, a significant technological change is required, in the first place the use of (more) mineral fertilizers. When the land is sensitive to erosion like is often the case in mountainous areas, the use of SWC structures is needed too; otherwise mineral fertilizers will simply be washed away (cf Kelly 2006). In the study area, construction of structural anti-erosion devices requires mostly labour. Fertilizer use also requires an increased use of labour as labour is needed to apply the fertilizer. Demand for weeding labour may also increase as fertilizers are likely to encourage the development of wild grasses and this may particularly be acute when soils are poor. It is a general complaint among farmers in the study area that fertilizers induce more weeds.

The decision problem the household must solve can then be formulated as :

$$\underset{c,s,F,g,L,l,x}{Max} \; u(c, g, L; J) \tag{6.1}$$

subject to:

| | |
|---|---|
| $c = q - s + F$ | staple consumption balance |
| $q = f(l, x; A)$ | farm technology |
| $(p_c + z) * F + p_q * g + p_x * x = y_0 + (p_c - z) * s$ | trade balance |
| $L + l = T$ | time constraint |
| $c{\geq}0, \; g{\geq}0, \; s{\geq}0, \; F{\geq}0, \; L{\geq}0, \; l{\geq}0$ and $x{\geq}0$ | non-negativity constraints |

where :
- $c$ = amount of staple consumed by the household
- $s$ = sales of the staple by the household
- $F$ = purchases of the staple by the household
- $g$ = purchases of non-staple by the household
- $L$ = leisure
- $q$ = household's production of the staple
- $x$ = amount of fertilizers applied
- $l$ = production labour
- $p_c$ = market price of the staple
- $p_x$ = market price of fertilizer

$p_g$ = market price of non-staple
$z$  = transaction costs per unit of the staple sold or purchased
$T$  = time endowment
$y_o$ = exogenous cash income (e.g. remittances, etc.)
$J$  = utility function shifters, e.g. the demographic structure of the household (household size, composition of the household population by age and sex), household tastes, initial wealth parameters, etc.
$A$  = production technology shifters, e.g. total cultivable land, number of workers, managerial skills of the household, inherent soil quality, climatic factors, etc.

The model comprises seven choice variables: $c, s, F, g, L, l$ and $x$. The predetermined (exogenous) parameters are $z, T, y_o, p_c, p_x, p_g$ as well as the production technology and utility shifters, $J$ and $A$.

A specific feature of the production function $f(.)$ is that it includes various farmer-based technologies that farmers in the study area use to fight erosion and possibly to improve yields, e.g., stone bunds, tie-ridging, manuring, etc. These technologies appear in the production process in the form of increased labour. Farmers do not construct new stone bunds in the study area; they are inherited from parents. However they require maintenance on a yearly basis. Maintenance is an agricultural activity distinct from the rest of activities. Tie-ridging are reconstructed completely every year but they are implemented simultaneously with tillage. Manuring is also often done with tillage or weeding.

There are two channels through which the farmer-based technologies could appear in $f(.)$. First, for most of them (e.g., tie-ridging, manuring) their level of application is captured by the variable $l$ (labour). Second, all of them contribute to change in soil quality. Hence we include these structural conservation technologies also as components of $A$ (production shifters).

*First-order conditions*

The Lagrange function associated with the problem can be written as

$$Max_{l,x,s,F,g} \quad u[f(l,x;A) - s + F, g, T - l; J] + \lambda[y_0 + (p_c - z)*s - (p_c + z)*F$$
$$- p_g*g - p_x*x]$$

If we add the non-negativity constraints on the choice variables $l, x, s, F$ and $g$ and solve for the optimum, we get the equilibrium condition for the model (the Kuhn-Tucker Conditions, or KTC):

$$u_c f_l - u_L \leq 0 \qquad l \geq 0 \qquad l[u_c f_l - u_L] = 0$$

$$u_c f_x - \lambda p_x \leq 0 \qquad x \geq 0 \qquad x[u_c f_x - \lambda p_x] = 0$$

$$-u_c + \lambda(p_c - z) \leq 0 \quad s \geq 0 \qquad s[-u_c + \lambda(p_c - z)] = 0 \qquad (6.2)$$

$$u_c - \lambda(p_c + z) \leq 0 \quad F \geq 0 \qquad F[u_c - \lambda(p_c + z)] = 0$$

$$u_g - \lambda p_g \leq 0 \qquad g \geq 0 \qquad g[u_g - \lambda p_g] = 0$$

$$y_0 + (p_c - z) * s - (p_c + z) * F - p_g * g - p_x * x = 0$$

The symbols $u_c$, $u_g$, $u_L$ and $\lambda$ are used to denote the marginal utilities of the staple, the non-staple, leisure and the cash income, respectively. $f_l$ and $f_x$ represent the marginal products of labour and fertilizer, respectively.

It can immediately be seen from the *marginal conditions* for $s$ and $F$ that the *non-negativity restrictions* for these two variables cannot both be non-binding. That is, at least one of these two marginal conditions must hold with strict inequality. Given the complementary conditions, this implies that either $s$ or $F$ is strictly positive or both may be equally zero at the farm household equilibrium. Indeed, if $s>0$ and $F>0$ it follows from the third and the fourth marginal conditions in (6.2) that:

$$\frac{u_c}{\lambda} = p_c - z \quad \text{and} \quad \frac{u_c}{\lambda} = p_c - z$$

which is clearly inconsistent.

Hence regarding the farm household participation in the market of the staple there are three possible regimes:

seller (s>0, F=0), or $\dfrac{u_c}{\lambda} = p_c - z$

buyer (s=0, F>0), or $\dfrac{u_c}{\lambda} = p_c + z$ $\qquad\qquad$ (6.3)

self-sufficient (s=F=0), or $p_c - z < \dfrac{u_c}{\lambda} < p_c + z$

The first-order conditions (equations system 6.2) suggest, however, that we need not worry only about change of regimes for the staple. We are confronted with a problem of regime switches for mineral fertilizer also.

Usually in the analysis of the comparative statics of the household model with data from poor countries there is more focus on the change of regime for the staple (equation 6.3). Here we focus on the regime switches for fertilizer. A consistent investigation of the comparative statics requires that all types of regime shifts that the model shows must be taken into account. If this is not done the benefits of public interventions, say provision of good roads to alleviate transaction costs, cannot be correctly assessed.

In our data set only one-third of the households in the study region apply fertilizers. Put another way, this means that, at the initial equilibrium solution, we expect to find often that the non-negativity constraint for fertilizer may become binding, i.e. $x$ takes a zero value or the household is a non-adopter of fertilizer. In response to a small perturbation, however, $x$ will become positive, that is the household becomes an adopter. The reverse applies for a household, which is set initially to be an adopter. Given the large number of non-adopters of fertilizers in the study area, switches from non-adoption to adoption and vice versa are indisputably to be considered as a primary characteristic for the type of household studied here. The specific feature of our model is that our representative household wants to adopt fertilizer but it often finds it difficult to finance it.[3]

**Table 6.1     Equilibrium of a semi-subsistence farm household: the alternative regimes**

|                   | Sales of the staple (s) | Purchases of the staple (F) | Fertilizers (x) |
|-------------------|:-----------------------:|:---------------------------:|:---------------:|
| Seller            |                         |                             |                 |
| Regime 1          | Positive                | 0                           | 0               |
| Regime 2          | Positive                | 0                           | Positive        |
| Buyer[1]          |                         |                             |                 |
| Regime 3          | 0                       | Positive                    | 0               |
| Regime 4          | 0                       | Positive                    | Positive        |
| self-sufficient   |                         |                             |                 |
| Regime 5          | 0                       | 0                           | Positive        |
| Regime 6          | 0                       | 0                           | 0               |

*1*     All households are required to purchase a minimum amount of consumer goods ($g$). Therefore, obviously, buyers and self-sufficient household must own exogenous cash income. They may however want to use part of this exogenous cash income to buy fertilizer.

---

3     For the rest of endogenous variables, $l$ (production labour) and $g$ (consumption of the non-staple), there is no doubt that the non-negativity restrictions will never bind (i.e. $l$ and $g$ are always strictly positive) at the solution. To restrict $l$ to be strictly positive is obvious. As for $g$, the restriction is acceptable as some minimum amount of non-staple (purchase of fish, kerosene, cooking oil, clothing, etc.) is required for the household to survive.

If we combine the two regimes for fertilizer (adoption versus non-adoption) with the three regimes for the staple (seller, buyer, self-sufficient) then we would arrive at a total of 6 (3*2) regimes for any level of the exogenous parameters as shown in Table 6.1. We are thus confronted with a multiplicity of transition points (i.e. points at which the solution mapping is not differentiable).

The simulation approach toward estimating models with multiple regime switches has been implemented with agricultural farm household data by Lopez (1984), with data from Canada, and Bhattacharyaa and Kumbhakar (1997), with data from India. The approach has some limitations. In limited samples it leads to what is known as the vanishing degrees of freedom problem (Johnston and Dinardo 1997, p. 305). Since exogenous variables are to come both from the consumption and the production sides of the household the number of parameters to be estimated can quickly become very large. As a result the model will be impractical if the sample size is small. To get rid of this problem Lopez used rather aggregated variables and in addition he set aside the variables for some inputs (e.g. the use of fertilizers) (Lopez 1984, p. 73). Hence one can suspect that Lopez's estimates suffer from omitted variable bias as input like fertilizers constitute an important instance of technology in agricultural production. In general, as we shall see in the previous chapter, when studying the agricultural production of poor countries some details about the inputs, the use of traditional structural soil conservation techniques and soil quality data are needed for the production technology to be estimated accurately, in particular when one adopts a cross-section estimation. If such details are to be maintained while the sample size is small, estimations based on simulation methods can indeed become quickly impractical.

We suggest a different estimation strategy that helps to circumvent the problem of vanishing degrees of freedom in samples with limited sizes. It may also show more consistently the effects of public interventions (e.g., reduction in transaction costs through road access) on the household input choices (in particular technology-enhancing inputs) in a context where agriculture is close to subsistence. We describe in the following section the general structure of the estimating strategy. More details about implementation issues are discussed in the next two sections.

*The estimating model*

At the core of multi-regime or non-separable household models is the idea that these models have a non-linear budget constraint. The idea is based on the assumption that in such models one or more prices are internally set by households (shadow prices), and these are used as decision prices by the households rather than the exogenous market prices. If we knew them we could use them in place of market prices, formulate a linear budget constraint based on these prices and solve the household decision problem with the classical optimization procedure. We cannot observe these prices directly, but they can be seen from the observed behaviour as in the optimum the relative shadow prices of the inputs are just their real marginal products in agricultural production.[4]

The relative shadow prices of the inputs will generally be different from their market counterparts. The gap between the market and the shadow prices can arise from various distortions affecting the markets on which the household operate. For labour inputs these may be labour market failures (e.g. missing labour market; high supervision costs, etc.); for fertilizer inputs, distortions will often be associated with liquidity constraints. Nevertheless, the argument we want to test here is that the gap arises mainly because of transaction costs in the market of the household output. If transaction costs in the output market are reduced, labour market failures and liquidity constraints for the purchase of fertilizers may substantially be reduced. If output is easily commercialized, this means higher monetization. It thus becomes easier to pay for wage labourers and mineral fertilizers.

Thus we estimate a system of equations where the gap between market and shadow prices for each of the inputs is linked to input use; within this same system a linkage will be established between these gaps and exogenous variables capturing various market distortions, focusing on variables which reflect transaction costs in the output market. Based on the estimated coefficients from such a system, we can simulate the changes in input use and the output change following a variation in the exogenous parameters (e.g., transaction costs).

The econometric model we solve is the structural equations system:

$$q = f(x, l, A, \beta) + \varepsilon \tag{6.4}$$

$$f_x = \tau_x \cdot p_x \tag{6.5}$$

$$f_l = \tau_l \cdot p_l$$

$$\tau_j = w_j(z, D), \text{ with } j = x, l, (x = \text{fertilizer}, l = \text{labour}) \tag{6.6}$$

where:

    $p_x$ is the ratio of the market prices of fertilizer and output;
    $p_l$ is the ratio of market wage and the price of the output;
    $f_x$ is the (real) marginal product estimated at the observed level of the

---

4   A similar argument was followed by Hall (1973). Jacoby (1993) has exploited Hall's idea to study the labour supply of peasants from Peru. Our approach differs from that of Jacoby in that we extend the analysis to peasant's demand of mineral fertilizer and the use of structural soil conservation techniques.

fertilizer input;

$f_l$ is the (real) marginal product estimated at the observed level of labour input;

$\tau_x$ and $\tau_l$ capture the gaps between the (real) marginal products and the relative market prices of the inputs;

$z$ captures the access to roads and

$D$ represents other variables explaining the degree of distortion.

Equation (6.4) shows the production function specification with functional form $f$. The production function is estimated in primal form and $q$ represents output in real terms (e.g., kg of staple). The equations (6.5) establish a relationship between the real marginal product (i.e. relative shadow prices) and the ratios of the market prices of the inputs and the output. In fact the equations denoted (6.5) indicate the first-order conditions of the usual profit maximization problem. The only (important) difference is that we incorporate into these first-order conditions factors $\tau_x$ and $\tau_l$. $\tau_x$ and $\tau_l$ are the distortion factors indicating the gaps between the market prices and the relative shadow prices (real marginal products). $\tau_x$ and $\tau_l$ capture transaction costs in the output market, but also other distortions affecting the market environment the household faces. Equation (6.6) gives a relationship between the distortion factor of each input and explanatory variables for these distortions. The variable for transaction costs is represented by $z$; $D$ is added for the rest of distortions (e.g., labour market failures, liquidity constraints, etc.). If we specify a change in $z$, we can use $w_j$ to simulate new values for $\tau_j$, which then appear on the right-hand side term in equation (6.5) to give new values for the marginal products of fertilizer and labour; from these we can derive new input and output levels. A complication arises for fertilizer. It is used only by 35 per cent of the households and on 16 per cent of the cultivated plots and we cannot, therefore establish a useful marginal product of fertilizer for these households. While a marginal product evaluated at zero could be used, this is only an upper bound to the marginal product that the household may see as relevant. Hence the approach is changed where fertilizer is concerned and a direct relationship between distance and its use is established.

The procedure has five steps:

a. Estimate econometrically a function that relates output in real term to the inputs (production technology) in the primal form; and derive the marginal product of labour at the observed level of its use.

b. Establish a relationship between each of the marginal products and the observed market price; this will lead to the computation of the price distortion factor, equal to the ratio of the marginal product of each input to the relative market price (the ratio of the market prices of the input and the output).

c. Establish econometrically a relationship between the distortion factor of each input and the exogenous parameters of the household model (including variable $z$) except the market prices; for fertilizer, establish a relationship between the input use and distance.

d. Use the estimates from the price distortion factor regressions to simulate new level of the marginal labour product or fertilizer use following a change in the variable $z$.

e. Simulate new input use and output level using the estimated new marginal labour product and fertilizer use. To do this the simulated (new) marginal products are inserted at the right-hand side of the first-order conditions shown by the equations denoted (6.5) above to infer the new input levels. Next, the new input levels are incorporated into the estimated production function (6.4) in the place of the observed values for the inputs to infer the new output level.

*Data*

The data are taken from the sample described in Adégbidi et al. (2004) (and in Gandonou and Oostendorp in this volume). They also derive the production functions that establish the first step in the procedure given above. Of interest for the analysis of the relationship with the market are the distances between the households and the nearest marketplace. For the four villages in the sample, these are given in the top row of Table 6.2. Okuaro is the village closest to the market, and Takuanta the most distant. These two villages also differ most in other respects. Takuanta lies in the mountains, and Okuaro mostly in the plains. Table 6.2 provides more detail on the differences between the villages and includes the means of the estimated values of the marginal labour products, based on the production functions from Gandonou and Oostendorp (this volume). It may be recalled that the production functions are estimated for cereals and beans only.

Table 6.2 shows that the more remote villages, Takuanta in particular, show higher degree of subsistence farming, slightly lower use of fertilizer and lower values for the estimated marginal labour products. Actual payments made to workers in working parties do not differ so much, however, probably because the food and drink involved are based on common traditions.

The estimated values of the marginal labour products are dramatically lower than the wage rates, and far below the national minimum wage rate. This even applies to the village closest to the market, Okuaro, though the difference is smaller.

## Estimation Results of the Effects of Distance

We investigate the effects of distance on the use of inputs in two ways. For labour input, we estimate how distance affects the value of the labour to the household, while for fertilizer, we estimate directly how distance affects its use.

**Table 6.2     Village characteristics 1998**

|  | Takuanta | Kutagu | Kunakogu | Okuaro | All |
|---|---|---|---|---|---|
| Travel time (minutes) | 122 | 96 | 107 | 27 | 88 |
| | | | | | |
| Total cash earnings hh (Fcfa) | 29,030 | 82,395 | 69,540 | 23,665 | 50,835 |
| Composition of earnings (%) | | | | | |
| Sales of crops | 4 | 23 | 43 | 38 | 30 |
| Sales of forest products | 27 | 0 | 21 | 0 | 11 |
| Sales of animal products | 4 | 18 | 5 | 17 | 12 |
| Wage labour | 3 | 0 | 0 | 0 | 0 |
| Seasonal migration | 0 | 3 | 0 | 18 | 3 |
| Self-employment | 50 | 25 | 17 | 27 | 26 |
| Remittances | 12 | 30 | 13 | 0 | 18 |
| | | | | | |
| % plots with steep slopes | 63 | 24 | 15 | 4 | 27 |
| % plots with stone bunds | 59 | 55 | 2 | 0 | 23 |
| % plots with tie-ridging | 34 | 29 | 91 | 15 | 39 |
| % households with fallow land | 80 | 64 | 29 | 32 | 52 |
| % households using fertilizer | 32 | 44 | 27 | 36 | 35 |
| | | | | | |
| % households with migrants | 13 | 72 | 16 | 4 | 28 |
| Annual pop growth 1979-92 (%) | -1.0 | -3.3 | -1.6 | 7.5 | 1.6 |
| | | | | | |
| Value marginal labour product* | 90 | 200 | 160 | 180 | 160 |
| Market 'wage' rate* (Fcfa/day) | 325 | 290 | 350 | 235 | 325 |

*Source* : NWO survey (1999); * median values
The market wage rates in the bottom row were derived from the in-kind payments that are usually provided to the traditional working parties. Note that the national minimum daily wage is Fcfa 1000.

*Labour*

For the first approach we use the marginal labour product, relative to the market wages as a dependent variable. This equals the variable $\tau_i$ in (6.5).

As the production functions were estimated at plot level, the values for $\tau_i$ are also at plot level, for those plots where cereals and/or beans are grown. The distance variables and other household characteristics are only known for the household. To use all information the estimating equation for $\tau_i$ is specified at plot level, and we account for intra-household correlation by specifying the plots as being clustered at household level.

The estimating equation for $\tau_i$ for household $k$ and plot $i$ is

$$\tau_{l,ik} = \phi_0 + \phi_1 z_k + \sum \phi_{2j} L D_{kj} \tag{6.7}$$

Here, $z$ is specified as travel time in hours, and the variables $D_{kj}$ are chosen so as to capture the various regimes as to being net buyers or sellers of food. These variables include household composition and farm size.

Village dummies were not included in the final estimation, as these turned out to take away the potential effect of the distance variable. Hence, the distance variable also captures possible village-specific elements to the extent that these are correlated with the variable.

**Table 6.3    Estimation of the effect of travel time on $\tau_l$ (ratio of real marginal product to real market wage)**

|  | Only travel time | Travel time farm, household size | Travel time and more household variables |
|---|---|---|---|
| Travel time | -0.15 | -0.15 | -0.18 |
|  | (-2.59)*** | (-2.61)*** | (-2.88)*** |
| Children, age 0–14 |  |  | 0.03 |
|  |  |  | (0.29) |
| Adults, age 15–59 |  |  | 0.10 |
|  |  |  | (0.79) |
| Household size |  | 0.003 | -0.05 |
|  |  | (0.21) | (-0.63) |
| Total land |  | 0.18 | 0.01 |
|  |  | (1.01) | (0.40) |
| Const. | 0.95 | 0.86 | 0.86 |
|  | (7.01)*** | (6.42)*** | (5.67)*** |
|  | $R^2 = 0.02$ | $R^2 = 0.03$ | $R^2 = 0.03$ |

Estimated with cluster effect and robust t-value in the parentheses; *** significant, 1% level.
N=260 plots

Results of the estimation are shown in Table 6.3.

Household specific variables turn out to be insignificant, but the effect of distance is to reduce the value of the marginal labour product. For every minute extra time, the relative marginal product falls by (0.15/60=) 0.0025. Thus, four minutes' extra time reduce household shadow daily wages by 1 per cent point of the ruling village 'wage', or by about Fcfa 3. The implication is that shadow wages are lower in more distant households. The lower shadow wages correlate with larger input of labour into the staple production and thus a higher production per ha.

*Fertilizer*

As discussed we estimate fertilizer use in a reduced form, skipping the intermediate role of market and shadow prices. This equation is estimated by interval regression, due to the limited number of values that the dependent variable takes. Next to travel time, we include a parsimonious list of control variables (which we believe are 'truly' exogenous). These include household size and total holding (total land area owned). We added also several variables describing the fixed plot characteristics: plot size, plot type, plot slope, (perceived) inherent plot fertility level. These variables are shown in Table 6.4.

**Table 6.4     Descriptive statistics for the variables for plot fixed characteristics**

| Variables | Obs. | Mean | Std. Dev. |
|---|---|---|---|
| Plot size (ha) | 260 | 0.6 | 0.49 |
| Plot type | | | |
| Rocky soil (dummy) | 260 | 0.22 | 0.41 |
| Gravely soil (dummy) | 260 | 0.44 | 0.50 |
| Plot slope | | | |
| light slope (dummy) | 260 | 0.36 | 0.48 |
| steep slope (dummy) | 260 | 0.17 | 0.38 |
| Distance from plot to home (km) | 260 | 0.56 | 0.88 |
| Perceived plot fertility (1 if fertility is relatively good)a | 260 | 0.46 | 0.50 |

ª The plot fertility variable is constructed based on a ranking of all the plots performed by each farmer. The variable included in the regression is a dummy which indicates whether the plot is ranked among the top half of the farmers' plots (cultivated and fallow plots) in terms of (perceived) plot quality.

*Estimation results and simulation*

Table 6.5 presents the econometric results obtained. The results are adjusted for clustering. The negative sign on the distance-to-plot variable means that plots close to the household's house *(tata)* are likely to receive (more) fertilizers. This corresponds exactly to what we have observed during the field research: plots close the *tata* are the most intensively farmed in the area. Note that these plots constitute a substantial part of the total land cultivated in the area (25 per cent of the total area).

The model indicates that travel time affects fertilizer use in two ways. Travel time to the plot reduces fertilizer use, but travel time from the *tata* to the market

**Table 6.5      Estimation of the effect of travel time on fertilizer use per plot**

|                                             | Coefficient | t-value   |
|---------------------------------------------|-------------|-----------|
| Travel time                                 | -3.68       | -2.03**   |
| Total land                                  | 1.37        | 2.74***   |
| Household size                              | 1.13        | 2.16**    |
| Distance from plot to home (km)             | -2.50       | -2.76***  |
| Light slope (dummy)                         | -2.44       | -0.66     |
| Steep slope (dummy)                         | -5.30       | -1.22     |
| Rocky soil (dummy)                          | 3.50        | 0.86      |
| Gravelly soil (dummy)                       | 5.16        | 1.33      |
| Perceived plot fertility (dummy, 1=good)    | -4.69       | -1.62*    |
| Constant                                    | 5.68        | 1.21      |

* significant 10% level; ** significant 5%; *** significant 1%;

reduces its use even more. A reduction of the travel time by one hour leads to an estimated 3.7 kg additional fertilizers.

Aside from the variables for distance from the plot to the house and for travel time, the variables for total land area owned, household size and perceived plot fertility are all significant. A test for joint significance of all the variables included into the regression except the variable for travel time showed joint significance (Chi-square 45.69).

The model results show that the effect of total land owned is positive. It may be hypothesized that it is the wealth effects which are captured here: a wealthier farmer may have better access to credit so that he can easily purchase fertilizer. We should recall here that most of the households which have used fertilizers on our sample have bought them on credit. The effect of household size is also positive. We suspect that it is the self-insurance issue which is revealed here. Larger households may be more inclined to intensify (e.g. by applying more fertilizers) in order to ensure that food needs are met. This result seems to be in line with the finding from Mulder (2000) in which it has been observed that subsistence farmers are prepared to apply fertilizers on basic food crops, although it is not profitable at the actual market prices of the output and the inputs. This result may also support one of the basic mechanisms in the Boserup's theory of intensification which assumes a positive relationship between population size (density) and the use of various intensive inputs including purchased inputs, e.g. mineral fertilizers (see Boserup 1965, Chapter 13).

To summarize, the results show that access to roads/markets matters for the improvement in fertilizer use. But it is not the only important factor. We found that for the same level of travel time, fertilizer use may vary. That is, there are additional factors that matter for improvements in fertilizer use.

*Simulation of effects of distance through labour and fertilizer use*

The significance of the estimated effects on production can be assessed by simulating the results of a hypothetical reduction in travel time to the market by 50 per cent. This leads to higher values for the shadow wages, as per Table 6.3, which would imply lower values for the input of labour; it also leads to more use of fertilizer, as per Table 6.5. In its estimated form, the production function of Gandonou and Oostendorp (this volume) leads to an expression for the marginal product of labour equal to:

$$Q_L = Q^* \cdot L^{\alpha-1} e^{bF},$$

where $Q^*$ stands for the effect of factors of production other than labour and fertilizer; L is labour input, and F is the input of fertilizer.

It is important to indicate here that the estimated production function includes only output from the staple food sector. The results thus show that farmers are prepared to shift labour from staple crop production to other uses when travel time reduces.

The microeconomic mechanisms behind these results can be traced out using different approaches. The mechanisms we suggest run as follows. Our theoretical set-up indicates that farmers' choices are based mainly on the real shadow prices of two variables:[5] real shadow price for labour and real shadow price for liquid funds. The latter price is important for the decision to apply mineral fertilizers. It is determined endogenously by confronting internal (household) supply and demand for liquid funds (see Petrick 2004 for a more formal model on this issue). The shadow prices cannot be observed but they can be derived as the real marginal products of labour and liquid funds.

When travel time is reduced, farmers are confronted with the following likely effects.

1.  Employment perspective becomes better in rural areas. Farmers are now able to adopt non-farm employment, both as workers and indirectly as producers and sellers of non-farm products. Because of these new opportunities for labour employment farmers have higher real shadow wages for labour and this leads to lower labour inputs in the farm sector.
2.  Lower travel time eases the interaction with the market, farmers have now less fear to sell staple food in exchange for, say, mineral fertilizers; the same staple food is bought more easily later from the market as markets are likely to be more integrated and the bands between buying and selling

---

5   In our theoretical set-up farmers are normally confronted with the shadow prices for three variables: labour, liquid funds and staple food crop. However we can set the staple food crop as the 'numéraire'. Then two decision prices remain: the real shadow prices for labour and for liquid funds.

prices are smaller. Higher sales means higher internal (household) supply of liquid funds. This may lower the shadow price of food (most households are net buyers) but also the shadow price of fertilizer. The latter prices fall as cash money becomes more easily available and travel costs fall.

3. Growing non-food crops for sales in the market become a more attractive option. This influences food production via its claims on land and labour. Thus, the shadow price of labour is likely to rise with a reduction in distance. As shadow food prices fall, real returns to staple crop labour will fall. The increased use of fertilizer may, however, counter this effect.

Focusing on food production, with $p.Q_L$ set equal to a higher shadow wage, $w^*$, L should fall, but the higher value of F, through $e^{bF}$, can compensate for this. In the simulation $w^*$ rises by 0.6 per cent in response to halving of the distance. This would normally reduce labour input by 1.3 per cent, but now we have the additional effect of F increasing by 2.7 kg or 27 per cent. The increased F has the partial effect of rising both Q and $Q_L$ by 1.2 per cent (F rises from 10 to 12.7; $e^{bF}$ rises from 1.047 to 1.060). In the end, labour input may even rise, and the higher use of fertilizer either mitigates or reverses the fall in labour input, and enhances production itself with the total effect ending up being positive.

It is interesting to make a distinction between the plots with and those without stone bunds. Production on cereal plots with stone bunds goes up by 2.9 per cent as a result of halving the distance from farm to market; production on cereal plots without stone bunds goes up by only 1.9 per cent. Thus, having the market closer by might mean that it becomes more attractive to use stone bunds. A test on the significance of this effect turned out to be non-decisive, however.

### Comparison with a Change in Population Density

The above derivation took a change in distance as point of departure. Halving the distance to the market makes labour more expensive, but stimulates fertilizer use. Eventually production will rise, even production of crops that are basically grown for subsistence. Reduction of the distance to the market has much stronger effects on other activities of the household. Households that live closer to the market grow more cash crops, and are more often involved in off-farm activities. Gandonou (2006) estimates the effects of a halving of the distance on the value of commercial crops to amount to Fcfa 9000 or 18 per cent of average cash income, while the probability of being in non-farm employment rises from 20 per cent to 26 per cent. Reduction in distance makes farming more remunerative, but also enhanced the returns to other activities. The effect of the former on investments in soil conservation could be positive, as the marginal value of stone bunds was shown to increase.

More distant households are therefore disadvantaged compared with households closer to the market. Can we compare this result with the effects of population density? Higher population density is assumed to lead to relatively higher (shadow) value of land relative to labour, and thereby to stronger incentives to invest in land quality. In this sense the effects of population density differ from the effects of distance. If markets come closer, the value of land may rise, as the more remunerative crops are grown, but the value of labour will also rise. If population increases the value of labour is bound to fall relative to land, the more so if the higher need for food puts further strain on production. The estimation underlying the above analysis of the effects of distance can also be used to trace the effects that larger household sizes would have. Larger household sizes, leaving the actual farm size unchanged, resemble increased population density.

We simulated the effects of population through the effects that household size has on the use of fertilizer (a significant effect of 1.13 kg/person in Table 6.5) and on the shadow wage rate (an insignificant 0.003 effect on the shadow/market wage ratio per person). A simulation in which household sizes were increased with one, two or three members (in three out of four villages[6]) led to the results shown in Table 6.6.

**Table 6.6     Effects of household size changes**

| Average household size change (%) | Output change (%) | |
|---|---|---|
| | with stone bunds | without stone bunds |
| 16 | 6.0 | 4.4 |
| 34 | 6.4 | 4.7 |
| 50 | 6.3 | 5.0 |

As Table 6.6 shows, production clearly increases with additional household size. This is due to demand effects (higher demand for food, higher shadow price of food, higher perceived marginal labour product and more use of fertilizer) and to supply effects (lower shadow price of labour, more supply of family labour). For a single extra person in every household production rises by approximately 8 per cent in case of stone bunds and almost 6 per cent on plots without stone bunds. The difference between the two indicates a rather weak incentive to use more stone bunds. Note that in both cases, the change in output does not meet the change in household size, so that food production per person would decrease.

---

6    Effectively, the first simulation amounted to adding one person to all the households, except in Kutagu; the second simulation added an extra person, except in Okuaro; and the third simulation added another person to all households except in Kunakogu.

## Conclusion

The estimation results confirm that the association between the wage distortion factor, and hence the shadow wage, and travel time is negative, as expected. Farmers in remote areas use a few, if any, external yield-enhancing inputs (e.g. mineral fertilizers); in replacement they use more labour in order to stabilize grain output, pushing down the marginal productivity of labour (i.e. the shadow wage). The simulation results show that the effects of a closer market on erosion control are marginal although it clearly induces higher incomes in Boukombé. The study has made it clear that reducing the distance to the market has slight, but positive, effects on the yields of grains. A more accessible market also provides households with more opportunities to grow other, more commercial, crops or to undertake other profitable activities.

An increase in household size also leads to higher production, and stronger incentives to invest in soil conservation measures, but the effects are very small. Labour productivity in this case is likely to fall. Out-migration on the contrary should increase per capita production substantially, but does not make SWC more appealing.

Nevertheless, the results of the study are encouraging. Even though they are isolated from major urban markets, farmers from the study region are responsive to market-based incentives (e.g. reduction of transaction costs, promotion of access to farm credit, etc.). Overall, the study results suggest that the region has reached a stage where further improvements in the agricultural performance and household welfare require less isolation, say by improving transport infrastructure to get the region connected to major urban areas. This also requires a further reduction of the credit rationing for the region as a whole. These two interventions are likely to reduce liquidity constraints to allow the use of more purchased inputs that can enhance labour productivity in the staple food sector; they can also lead to an expansion of the set of income-generating activities with higher labour productivity (cultivation of high-value crops, rural non-farm sector). The study results suggest that the response of Boukombé farmers to such actions can be positive and substantial.

Chapter 7

# Agricultural Intensification in the Koza Plain Drylands, Cameroon: Ongoing Trends and Possible Futures

Wouter T. de Groot and Adri B. Zuiderwijk

## Introduction

This chapter focuses on the Koza plain, a dryland area in north Cameroon that has been populated rapidly during the last decennia by farmers of the Mafa ethnic group who descended from the adjacent Mandara mountains, where they practise an intensive, highly integrated, terraced subsistence farming system (Zuiderwijk 1998; De Groot 1999a; Seignobos and Iyébi-Mandjek 2000). With rapidly growing population density and the onset of soil degradation, intensification of agriculture in the plains has become unavoidable. As we will see below, many farmers are taking measures to maintain soil fertility, but the trend of soil degradation does not appear to be halted yet. This raises the question of what the future of the Koza plain, if any, may be.

In order to address this issue, theoretical perspectives are of great importance. They will be explored in the second section, after a brief description of the research area given in the previous section. The third section will focus on intensification trends, which then will be interpreted in the next section.

## The Study Area

The Mandara mountains form a chain of deeply dissected hills which straddle the border between north Cameroon and north Nigeria. The region is characterized by highly seasonal rainfall which is erratic in its distribution over time and space and by generally thin and infertile soils. The mountains have long been a refuge area to escape from the pressure of Islamic kingdoms which ruled the lowlands, and thereby became one of the most populous, ethnically diverse and culturally isolated regions in Africa. The Mandara mountains are inhabited by 23 different ethnic groups. The Mafa, being the most numerous one, occupy the central and eastern part of the northern offshoot of the Mandara mountains, an area of about 1500 km². Population densities run up to 350 inhabitants per km².

Mafa agriculture in the mountains is the product of ages of agricultural intensification, spurred by population growth in a context of limited possibilities for territorial expansion. Until about sixty years ago, Mafa territory was limited to the mountains proper. Mafa communities lived in almost complete isolation from each other and from the ethnic groups occupying the sparsely populated surrounding plains. In the isolation of the mountains, the Mafa practised a permanent and labour-intensive form of subsistence agriculture.

Traditional Mafa farming is based on a wide variety of soil management techniques. The hillsides are covered with constructed terraces that have reached a state of exceptional perfection. Other ethno-engineering techniques involve small-scale irrigation, canalization and drainage systems. In addition, farmers in the mountains practise a wide range of soil fertility management techniques, including crop rotation and mixed cropping, agro-forestry, and biomass and nutrient management. The two staple crops, sorghum and millet, are rotated every year by all farmers in unison, in order to prevent the build-up of crop parasites such as *striga*, and are grown in close association with other crops such as groundnuts, sesame and beans that all have their own small-scale niches on the terrace. Trees are also an important element of the Mafa land use system. They fulfil economic functions such as firewood provision as well as environmental functions such as the stabilization of terraces and the fertilization of the soils (as a result of their deep roots and production of organic litter). Moreover, an intensive livestock-raising system is a central element in soil fertility management. Livestock includes smallstock and a limited amount of cattle. In the dry season between December and May, livestock is allowed to roam free in order to consume crop residues and leaves of wild bushes. During the agricultural season, livestock is penned and hand-fed in the family compound. The manure that accumulates in the stables is collected, stored and eventually spread out in the fields at the end of the dry season. The intensity and ingenuity of Mafa nutrient management is illustrated by the fact that termites are used to digest crop residues and then fed to the chickens.

Many aspects of traditional Mafa institutions can be traced back to the material facts just described. Political power and authority in Mafa society is limited and dispersed. The individual household, the *gay*, forms the economic nucleus of Mafa society and hardly any authority exists above that level. The *gay* is a kinship group, a virilocal residence group and the basic unit of production and consumption, all at the same time. The household almost exclusively supplies agricultural labour. Although some forms of labour exchange between households and communal labour arrangements exist, their economic significance is limited and they generally concern non-agricultural activities. Task division along age and sex lines within the *gay* exists only to a limited extent, a feature that the Mafa share with other intensive mountain agriculturists such as the Kofyar (Stone et al. 1991). In the mountains the compounds are scattered over the hillsides with a distance of about 50 m or 100 m between them. Household plots generally encircle the compound. The vicinity of the compound to the plots enables the constant care and attention needed in intensive agriculture. Land is subject to a system of informal but clear-cut private property rights and transmission mechanisms. This system is the central condition for the intensity and permanency of traditional Mafa agriculture (Martin 1970). At the same time, it is at the basis of the independent position of the *gay*.

In the course of the second half of the twentieth century, many mountain dwellers moved downhill to the Koza plain. Push factors were the population growth, land scarcity and lack of drinking water in the mountains. Pull factors were the availability of land, medical services and, especially, potable water in the plains. In addition to this spontaneous migration, organized migration took place as well. So-called *casiers de colonisation* (settlement schemes) were put into place during the 1950s, by opening up sparsely populated areas with large-scale clearing of land and rectangular road lay-outs. In *casiers* such as the one in the Koza plain, plots of 4 to 9 ha were distributed among the migrants. They did not become the owners of these plots but received the use rights for an indefinite period of time.

In the plain, migrant farmers were encouraged to grow cotton. Cotton cultivation took place under the close supervision of the marketing board CFDT and, since the 1970s, SODECOTON. The plains of Koza became one of the first and initially one of the most innovative cotton-growing regions of north Cameroon. The cultivation of cotton became the vehicle for innovations such as plough agriculture, new cropping patterns, improved seed and later the use of (subsidized) industrial inputs, essentially fertilizers and chemical pesticides. In the 1960s, per ha productivity of seed-cotton in the Koza plains was among the highest of all sectors of north Cameroon, as was the use of mechanized equipment and fertilizers (Zuiderwijk 1995, 1998).

During the 1960s and 1970s an extensive land use system was used in the plains, which contrasted strongly with traditional cultivation that the farmers had practised in the mountains. The new and less constraining physical conditions as well as the new social and institutional conditions in the plains appeared to allow for a different and less prudent type of farming. The introduction of the plough provided farmers with the possibility to increase the area under (cotton) cultivation per household. Traditional fertility management techniques were largely abandoned because they could be replaced by bush fallow and industrial fertilizers. Traditional anti-erosion measures were abandoned and trees were eradicated because they interfered with mechanization.

The agricultural system that was used in the plain has often been described as an initially rational response to decreasing man/land ratios but increasingly putting livelihood at risk (Martin 1970; Boulet 1971; Campbell 1981). As long as there was still virgin land to clear, the bush fallow system offered maximum returns to agricultural labour. In the course of the 1980s the plains became saturated, however, and in most areas of the plain, bush fallow was no longer possible. Annual cultivation became common practice in the Koza plain. This shift to annual cultivation was generally not accompanied by adequate measures to maintain the natural capital, i.e. to restore soil fertility. In the 1990s soil fertility decline had become a widespread problem, further aggravated by rather abrupt rises of the price of industrial fertilizer, first due to the termination of subsidies in 1987, and later because of the devaluation of the local currency (Fcfa) by 50 per cent in

1994 (Zuiderwijk and Schaafsma 1997). The increasing scarcity of bush and free pastures contributed to a decline of livestock numbers in the area, further limiting the availability of nutrients. Food production has stagnated and many farmers complain that they have to work more and more hours on fields that produce less and less. Most farmers nowadays spend several weeks to several months a year on migration in order to supplement on-farm income. The possibilities for remunerative employment in the urban sector are very limited in Cameroon, however. Obviously then, things will have to change in Koza plain land use. This will be the subject of the next sections.

### Theoretical Considerations

An optimistic perspective on the relationship between population growth and agricultural intensification is given by the neo-Boserupians who expand on the seminal work of Ester Boserup concerning agricultural intensification. While Boserup herself maintained that population-driven intensification goes together with declining returns to labour, the neo-Boserupians assert that under the right conditions, agricultural intensification can go hand in hand with improved rural welfare and prevention of land degradation (Turner et al. 1993; Tiffen et al. 1994; Templeton and Scherr 1997). Tiffen et al. (1994) stated for instance, that:

> Conservation pays when land becomes scarce under existing technological conditions [...] The growth of population, working conjointly through an increase in the labour force and the growth of markets, has driven intensification on the smallholdings [...] This intensification has characteristically taken the form of investment in sustainable technologies and management. Investment funds transferred from other sectors have themselves been generated by a diversification of incomes, made possible –and necessary – by high household fertility. (pp. 28–9)

Tiffen's historical and regional approach is a valuable one, as it calls attention to long-term trends of (semi-)autonomous agricultural change. However, as we have argued elsewhere (Zuiderwijk 1997), the evidence presented on the precise relationships between key variables such as population, markets, social differentiation, investment and productivity, is weak. As has been shown by Murton (1999), these shortcomings may lead to wrong conclusions about the extent and mechanisms of agricultural intensification. Murton found that farmers in Machakos District who intensified successfully were generally those with access to remunerative non-farm income. Poorer farmers, dependent on local agricultural employment, struggled to keep their farms in shape or ended up as labourers making terraces on the land they once owned. As Murton (1999, p. 40) puts it, 'Boserupian intensification on richer farms, and a form of Geertzian involution on poorer farms are seen to be proceeding side by side within the same village'.

Murton here makes use of a dichotomy of two types of intensification which are found in the land use literature more frequently, and that may be summarized as follows:

- On the one hand, there is the type emphasized by the neo-Boserupians, which may be called agricultural *transition*. It is characterized by farmers investing in innovation and in the quality of their land. Returns to labour may be depressed during the investment period, but may be good again when the new systems (trees, terraces etc.) are in place.
- On the other hand, there is the ('Malthusian') type emphasized by Geertz, and that may be called agricultural *involution*. It is characterized by farmers continuing with the same farming system they had before, only replacing the once free fertility maintenance services of nature (through fallows and other bushland functions) by putting in more and more labour per hectare. Soil fertility may continue to decline, however, and returns to labour decline with it.

Transition and involution may follow each other in time. In the humid tropics, for instance, a forest-based slash-and-burn system may involute towards ever shorter fallows but may then be followed by farmers going into transition and constructing systems for irrigated, terraced rice. This system may then involute again, with farmers working ever harder on ever smaller plots.

As Murton (1999) and De Groot (1999b) indicate, transition and involution may also be seen as two pathways running parallel to each other in the same place, bifurcating from a common starting point. This starting point may be an involuting extensive system, such as the bush-based system of the Koza plain and of the Sahel in general. In one area or one type of farm, farmers may have both the capacity (capital) and the motivation (incentives) to innovate and invest in the land, so as to escape from the prospect of degradation and dwindling income. In a different area or a different type of farm, farmers either may be not motivated (yet) or may have become so poor already that they lack the capital to invest (the 'poverty trap'). These farmers or areas may then be stuck in an involution pathway, which may either level off in a sustainable but low returns-to-labour scenario, or continue to be unsustainable and sink ever deeper into poverty, landlessness and out-migration.

A second set of theoretical considerations concerns the question of what could be the *determinants* of the farmers' choices to go either the one way or the other. The first thing we note then is that external markets rather than internal (population) factors will often play a crucial role. The miracle of Machakos, for instance, may well have been brought about primarily not by that the coffee market was booming in the 1970s and that coffee in Machakos happens to require terracing due to the sub-humid climate, coupled with the fact that in the same period, Machakos found itself in the expanding horticultural zone around the rapidly growing city of Nairobi. These lucky circumstances my have helped much to generate both the motivations and the capital needed for the transition. In the same urban-based vein, Murton (1999) found that much of Machakos' investment capital had in fact been earned in the city. Retiring civil servants, for instance, were often the forerunners of the transition, and often the greatest beneficiaries. Besides these economic factors, knowledge and agricultural extension may play an important role too, not only by supplying technical know-how but maybe even more importantly, by raising awareness that innovation is necessary and that farmers should do it timely, before they are caught in the poverty trap. And finally, cultural factors may exert an influence, too. The Kamba of Machakos are well known for their industrious nature, for instance, and for the capacity of women to take decisions also if the husbands are only interested in non-agricultural pursuits or absent for long periods of urban migration. Factors such as these may well play a role alongside with, or possibly even dominating, population-density-dependent factors such as land scarcity, labour availability and the efficient circulation of innovative ideas.

A last theoretical consideration points at the soils rather than at the people. Good soils may respond much faster than poor soils to improved management practices and generate higher returns to labour, thus liberating capital and energy for further improvements. A second element in this area regards the pivotal role of organic matter in soil productivity. Organic matter when decomposing delivers nutrients to the soil but also the carbon-rich particles that keep these nutrients, and those of fertilizers, in their place and available for plants. Moreover, organic matter sticks the fine particles of the African soils together to form the 'macro-aggregates' that give the soil an open structure that allows rainwater to infiltrate. And finally, organic matter greatly increases the water-holding capacity of the soils, so that infiltrated water will remain available for plant growth. Organic matter management is the key of the sustainability of extensive farming systems and intensive farming systems alike. In these terms, the transition from sustainable extensive systems to sustainable intensive systems may be described as the transition from *ex situ* organic matter subsidies of the arable land (basically, by means of using the bush for cattle and fallows) to *in situ* organic matter cycling on the arable land itself, e.g. by means of penned cattle, crop residues, green manure, on-farm trees, composting or, at a larger spatial scale, using organic urban waste or biomass subsidies from good soils (such as heavy vertisols) to poor soils. Thus for the Koza plain as any other place, the pivotal question for transition becomes where the organic matter should come from.

**Intensification in the Koza Plains**

In this section we will present results from a study on agricultural intensification in the Koza plain, gathered largely in the framework of a wider project called 'The Transition of Tropical Land Use', financed by NWO in the Netherlands and also involving Kenya, Benin and the Philippines. Data were collected between 1996 and 1999 in four villages in the plain: Ziler, Djinglia, Malta Maya and Ouagza

Gabas. Djinglia lies close to the mountains and has many gently sloping soils, which were also found in parts of Malta Maya. Ziler and Djinglia are densely populated. Population densities are approximately 250–300 inhabitants per km² and the size of most holdings does not exceed 1 hectare. Malta Maya and Ouagza Gabas are less densely populated. Here population densities are in the order of 150 to 200 inhabitants per km². The size of most holdings is about 2 to 3 hectares. The study included village studies and household interviews.

**Table 7.1**     **Summary statistics of 508 plots, surveyed among 99 households in Cameroon**

| Use of plots (percentages) | | | Tenure situation (percentages) | | | |
|---|---|---|---|---|---|---|
| Grazing | Fallow | Crops | Private titled | Traditional private rights | Rented in / share cropping | Other / unknown |
| 0 | 0.83 | 99.17 | 27.57 | 29.54 | 33.92 | 8.97 |

| Slope types (percentages) | | | Annual Crops (%) | | | |
|---|---|---|---|---|---|---|
| Low flat | Lower slope | Very steep | Cereal | Vegetables | Leguminous | Other cash |
| 28.57 | 67.86 | 3.57 | 21.85 | 27.36 | 1.77 | 45.87 |

The average household size in the sample is 7.9 members, of whom 42 per cent is below the age of 15. The average household has about 5 plots of land, with an average size per plot of 0.56 ha. About 31 per cent of the households are engaged in non-farm self-employment, and 6 per cent in wage employment.

In the course of three decades annual cultivation has become the norm in the Koza plain. Years of continuous cultivation have taken place and soil degradation has become recognized as a major problem both by the farmers themselves and by extension agents active in the region. SODECOTON, the most important extension agent in the area, becomes increasingly involved in natural resource management. It (re-)introduces anti-erosion and soil fertility management techniques, such as the 'sowing of fertilizers', near-contour ploughing and the use of animal manure. Also because of environmental considerations, SODECOTON recently lifted an old regulation concerning the uprooting and burning of the old cotton stand immediately after the harvest. It also lowered the price of fertilizers a little, which amounted to some subsidization, according to local SODECOTON officials. What about the farmers themselves? Farmers' responses to the need of intensification and the problem of soil degradation are varied, but some trends are discernible.

*Cotton cultivation*

The first trend in cotton cultivation is its increasing total acreage. Since the 1950s the area cropped with cotton in the Koza plains steadily increased, both absolutely and as a percentage of total cropped area. This general trend (despite some decline in the early 1970s and early 1990s) accelerated during the late 1990s. Since the devaluation of the Fcfa (by 50 per cent) in January 1994, which resulted in higher farm-gate prices for seed-cotton, the acreage cropped with cotton rose very sharply by more than 50 per cent. In the course of the late 1990s, cotton cultivation even expanded into the mountains, increasingly replacing millet in the dominant sorghum–millet rotation cycle (De Groot 1999a).

Cotton produces high yields in monetary terms, about 50 to 70 per cent higher than those of sorghum, but also demands much more labour (about 50 per cent more) and external inputs. In the course of time, cotton growing became more and more input intensive. Here we present some data (referring to 1996/97) on the intensity of cotton cultivation in the Koza plain.

Virtually all cotton cultivation makes use of chemical inputs delivered by SODECOTON and is supervised by SODECOTON. On the total of 5041 ha cultivated in the 'section Koza', which corresponds more or less with the Koza plain, only 33 ha was considered to be traditionally cultivated, i.e. without the supervision and support of SODECOTON. In the neighbouring sections of Mora and Dogba, this percentage of traditionally cultivated cotton was 12 and 17, respectively. In the Koza section draught cultivation is widely used in cotton. In 1996/97 nearly 47 per cent of the fields were ploughed by oxen; 50 per cent of the fields were prepared manually; 3 per cent was non-tilled, and tractors were used on less than half a per cent. Rotation was strictly adhered to. In Koza the preceding crop was never cotton; on 93 per cent of the acreage it was a cereal, and on 6 per cent it was groundnuts (which are considered a good precessor from an environmental point of view because of the plant's ability to fix nitrogen). Note that this implies that the land is never fallowed. Herbicides (which act as labour-savers) were used on only on 4 per cent of the acreage. Fertilizers however were applied on 96 per cent of the acreage. They were sowed or directly covered while hilling up; nothing was spread out just like that. The heavy use of external inputs was reflected in the amount of agricultural credit supplied by SODECOTON. In Koza section, this amounted to 193 million Fcfa, or about 24.500 Fcfa (37 euro) per grower.

In the Koza section the social distribution of cotton production was as follows. In 1997/98, the 5009 ha of (intensive) cotton was cultivated by 8068 growers (0.63 ha per grower on the average). About 66 per cent of these growers cultivate 0.5 ha or less with cotton, 23 per cent cultivate 0.75 to 1 ha, 9 per cent cultivate 1.25 to 2 ha, and 2 per cent cultivate larger areas.

In the two densely populated villages, cotton was a smallholder's cash crop. Most growers cultivated less than one *quart* (0.25 ha) of cotton. They cultivated it in a very labour-intensive way and applied as much manure as possible. Labour input was about 70 days per *quart*. Nearly 20 per cent of labour input was in field preparation, manuring and tillage. Sowing and harvesting amounted to about 45 per cent of total labour input. The remainder was in weeding. In the two less densely populated villages, a limited number of commercial farmers dominated cotton cultivation. They grew cotton on larger tracts of land (each about 1 to 2 hectares). Tillage and weeding were mechanized; labour input was about 53 days per *quart*. Labour input in field preparation, manuring and tillage together only amounted to about 10 per cent of total labour input. Labour input in sowing and harvesting together amounted to nearly 60 per cent of total labour input. These farmers did not apply manure but used large quantities of industrial fertilizer.

Hence, two systems of cotton cultivation appeared to exist in the Koza plain:

1. One was based on industrial fertilizers and mechanization. Here cultivation was relatively large scale. Most labour was put in sowing and harvesting. Labour input in field preparation was limited.
2. Another is based on high use of labour and organic fertilizers. Here cultivation was relatively small-scale and intensive. Field preparation took nearly three times as much labour as under the first system.

Yields did no differ significantly between the two systems, but because of differing production scales and labour inputs, the returns to labour and the incomes derived from the cultivation did. Farmers who owned oxen and ploughs, and who were able to mobilize the necessary resources to apply inputs and labour in time, were able to realize high monetary returns. Farmers who did not were obliged to work long hours at relatively low returns to labour.

In the Koza plains as a whole, yields slowly but steadily increased during the first two decades of cotton cultivation (1960s and 70s) and stabilized at about 1100 to 1200 kg per ha during the 1980s and 1990s. According to the interviewed farmers, this levelling-off of yields despite increasing labour inputs and increasing sophistication of cultivation practices was caused by soil fertility decline and increasingly unstable rainfall.

*Maize cultivation*

Another trend in the Koza plain was the increasing cultivation of maize, partly replacing millet (and to a lesser extent sorghum) as a staple crop. Maize has been assigned an important role in the intensification of agriculture in Nigeria and Kenya (e.g. Ade Freeman and Smith 1996), but in the Koza plain, maize cultivation was as yet limited to only 5 to 10 per cent of the total acreage. This may be connected to the fact that maize was largely confined to the fertile clay soils near the rivers and the plots next to the compounds, where farmers generally put their organic household waste.

The cultivation of maize generally involves the use of large quantities of industrial fertilizers. If conditions (soil, rainfall, fertilizers) are good, maize can produce a good yield, about 20 to 30 per cent higher than sorghum. Maize cultivation also demands somewhat more labour than sorghum cultivation, however. The IRA, the national agricultural research agency, had introduced short-cycle varieties of maize. They provided farmers with the opportunity to sow maize when the first sorghum plantings had failed due to shortages of rainfall during May or June, at the beginning of the rainy season.

Maize cultivation was especially popular in Ziler, a densely populated village with relatively good *drob* (clay) soils. Here about one fifth of total acreage was covered with maize, producing (in monetary terms) about 30 per cent of total agricultural output during the rainy season. Most maize was sold in order to buy the culturally preferred sorghum. Maize was also grown in Malta Maya and, to a lesser extent, in Djinglia and Ouagza Gabas. The differentiation of labour intensity described for cotton was present also for maize. In Ziler and Djinglia maize was cultivated in a more labour-intensive way, with input per *quart* of 49 days. In Malta Maya and Ouagza Gabas, labour input was only 35 days per *quart*.

*Onion cultivation*

By far the most important innovation in the Koza plains was the cultivation of onions. Onions have been cultivated in the Koza plains for decades, but they assumed major economic importance only in the 1990s. Nowadays, the income derived from onions makes up a substantial share of total agricultural production. The produce is largely destined for urban markets (Douala, Yaoundé) in the south of Cameroon.

Onion cultivation takes place during the dry season on plots with high clay content, which are situated near the wet-season streams (*mayos*). Cultivated plots were small, about 0.7 *quarts* on average, with a range from 0.25 to 2 *quarts*. These plots were subdivided into rectangles of about 2.5 m by 1.5 m, locally called *carrous* in French or *ntaya* in Mafa. These rectangles were connected by small irrigation channels to a water source, generally a hand-dug well, from which irrigation water was pumped up from the groundwater table, which is generally at a depth of about 2 to 3 m (in December) or 5 to 10 m (in March/April). This variation proves that the groundwater is replenished every year by infiltration from the *mayo*. A long-term trend (i.e. the sustainability of water extraction) is not known.

Onion cultivation was very labour intensive. On average it took about 150 days to cultivate one *quart* of onions, which is about two to three times as much as the labour input in rainy season crops. Labour expenditure was about equally divided between field preparation including the construction of a well and the construction of *ntaya* (24 per cent), sowing/transplanting (22 per cent), weeding (20 per cent) and harvesting (29 per cent). Much more than was the case in rainy season agriculture, use was made of non-family labour such as wage labour and 'help' labour. Wage labour covered on the average about 40 per cent of total labour input and was employed in all of the four above-mentioned agricultural activities, but especially in the construction of *ntoya* and in weeding. On the average about 20,000 Fcfa per *quart* was spent on wage labour. Cash expenditures on inputs (fuel, fertilizers and insecticides) were even higher, about 90,000 Fcfa per *quart*. Most cash was expended on fuel. About half of the cultivators received their inputs on credit.

Onions produced high yields and growing them could be a very profitable activity. Our sample of onion cultivators (n = 28) realized yields ranging from 1800 to 7500 kg per *quart*. The sample included farmers who realized net incomes from onion cultivation of about 600,000 to 1 million Fcfa (900 to 1500 euros). This amounted to about six times the average yearly household income in the area. These farmers cultivated 0.5 ha, with yields of about 10 to 12 metric tons. Most growers, however, earned about 50,000 to 150,000 Fcfa, while some growers even ended up with a deficit or only a small income of 10,000 to 20,000 Fcfa.

Naturally income from onion cultivation is largely a function of production scale and yields realized. But much depended also on the price the farmers were able to get for his produce. The onion market was characterized by extremely high price fluctuations. Prices were lowest (about 30 to 50 Fcfa/kg) just after the harvest in March and April, and highest (about 150 to 250 Fcfa/kg) at the end of the rainy season in October and November. Stocking onions could therefore be quite profitable, but it is difficult to do this during the rainy season. The prices received by the farmers in our sample ranged from 40 to 130 Fcfa per kg, with a mean of 85. Small farmers who cultivate onions at a scale of less than 0.5 *quart* typically got lower prices, generally in the order of 40 to 60 Fcfa per kg, because many small growers got their inputs on credit, which forced them to sell their produce early. Large farmers generally received higher prices, ranging between 80 and 100 Fcfa. This multiplication effect of acreage and price is one of the reasons why to a much greater extent than in rainy season agriculture, incomes from onion cultivation largely accrued to a limited number of relatively rich households. The increase of onion cultivation was therefore paralleled by an increase of inequality. Furthermore, as the proceeds from onion cultivation were often invested into arable land expansion, onion cultivation was leading to an increasing concentration of agricultural land.

Onion cultivation was very limited in Djinglia, but widespread in the other three research villages. This was especially so in Ziler and Malta Maya, the villages located adjacent to major *mayos*. In Malta Maya the gross agricultural product realized in onion production almost equalled that of total rainy season agriculture. To a lesser extent this also goes for Ziler. Given the high input costs (especially fertilizers and fuel) involved in onion cultivation, this was less outspoken for the net agricultural incomes but still quite substantial. Added to this monetary importance is the fact that onion cultivation employed large amounts of labour in the dry season, when other local opportunities were scarce. Thus, onions could help to supply capital needed to invest for agricultural transition also on the non-onion land. One major setback of the onion cultivation in this respect is that nonresidents, e.g. traders from Maroua, controlled many of the most profitable onion plots; re-investment of profits is then not likely to take place locally.

*Indigenous intensification techniques*

Apart from the introduction or increasing cultivation of modern high-input crops, many farmers appeared to apply intensification techniques in existing crops, some of which were traditionally applied in the mountains.

Multi-cropping, the most important traditional intensification technique, was generally practised by farmers who disposed of only limited and marginal land. Usually they cultivated a number of small plots in the foothills, which were mainly composed of *wuyak* (sandy) soils. This is especially so in the foothill villages of Djinglia and part of Malta Maya. Multi-cropping here aims to stabilize and to maximize the yield from a given plot. However, since multi-cropping is generally incompatible with mechanization, it demands much more labour than mono-cropping. Multi-cropping therefore typically combines high production per hectare with low returns to labour

Most of the interviewed small farmers used a wide range of organic fertilizers, ranging from animal manure to household debris and green manure. Increasing the recycling rate of organic matter is a second 'indigenous' way to intensify agriculture. In villages such as Djinglia and Ziler, stall-feeding on crop residues had largely replaced grazing. In parts of Malta Maya, and to a lesser extent Ouagza Gabas, this process was underway. The quantities of manure and other organic matter sources were generally low, however, and the possibilities for increasing these quantities were limited. Nearly all biomass produced in the foothills was already used one way or the other. Using groundnut shells as a cattle feed, for instance, necessarily reduced their availability as green manure. According to farmers and statistics provided by the veterinary services, the number of cattle and small livestock in the Koza plains has actually decreased during the early 1990s, either as a result of harvest failures or as a result of increasing scarcity of grazing land (bush).

A third set of indigenous techniques for intensification applied by the farmers concerned investments in anti-erosion measures. An example of these is the *ingaleam*, or 'water barrier'. The *ingaleam* is generally constructed of earth, fortified with twigs and sometimes stones. It acts to retard run-off flow, as a silt trap, and as a kind of compost heap at the end of the season. Generally *ingaleams* have to be repaired or reconstructed every year. Another example is the *ndaldar*, a kind of living fence, serving primarily to keep cattle and small-stock out of a field, but also to produce biomass and to halt erosion. About 10 per cent of all plots have contour bands. Tables 7.2 and 7.3 show the extent to which such plots and households differ from the rest.

The contour bands, of course typically used on sloping plots, also show more use for cash crops, including mixed cropping (here counted double). These plots

**Table 7.2     Household characteristics by contour bands**

| Investments in Contour bands | | No | Yes |
|---|---|---|---|
| Number of plots | | 459 | 49 |
| Average size of household | | 8.76 | 8.65 |
| Male head of household (percentages) | | 87.8 | 93.88 |
| Age head of household | | 43.68 | 43.56 |
| Household members aged 0 to 15 (percentages) | | 43.53 | 44.46 |
| Household members aged 16 to 59 (percentages) | | 51.28 | 45.31 |
| Household members aged 60 and older (percentages) | | 5.20 | 10.23 |
| Average number of people per household with: | No education | 3.94 | 4.29 |
| | Primary education (incl. adult literacy) | 2.56 | 2.20 |
| | Secondary education (incl. intermediate) | 0.46 | 0.37 |
| | Higher education (university/ college) | 0.00 | 0.00 |
| | Other education / Unknown | 0.00 | 0.02 |
| Avg. number of plots per household | | 5.97 | 5.35 |
| Migrated household member (percentages) | | 8.30 | 8.62 |
| Remittance in local currency (average/month) | | 2120.92 | 2642.86 |
| Cash expenditure quartiles (percentages) | 1 (poorest) | 23.39 | 28.57 |
| | 2 | 25.17 | 32.65 |
| | 3 | 25.17 | 24.49 |
| | 4 (richest) | 26.28 | 14.29 |
| Households engaged in off-farm labour (percentage) | Wage employment | 5.19 | 4.13 |
| | Non-farm self employment | 30.38 | 31.3 |

are closer to the house. Households using contour bands appear to have more dependent family members, and show lower cash expenditures.

**Interpretations**

In order to arrive at an insight in the future of the Koza plain we first take a look at the various crops separately. Cotton appears to be part of an unsustainable Malthusian pathway. Theoretically, cotton can be grown sustainably in an extensive bush fallow system but, as we have seen, such a system is not applied any more even in the two less densely populated villages studied. There is simply

**Table 7.3     Plot characteristics by contour bands**

| Investments in Contour bands | | No | Yes |
|---|---|---|---|
| Number of plots | | 459 | 49 |
| Distance to home (meters) | | 2143.19 | 1260.00 |
| Average size of plots (hectares) | | 0.52 | 0.79 |
| Use of plots (percentages): | Grazing | 0.00 | 0.00 |
| | Fallow | 0.95 | 0.00 |
| | Crops | 99.05 | 100.00 |
| Tenure situation (percentages) | Private titled | 27.60 | 27.27 |
| | Traditional private rights | 29.06 | 34.09 |
| | Rented in / share cropping | 34.14 | 31.82 |
| | Unofficial, squatting, still obtaining title deed | 0.00 | 0.00 |
| | Other / unknown | 9.20 | 6.82 |
| Slope type (percentages) | Low flat | 31.25 | 9.09 |
| | Lower slope | 65.94 | 81.82 |
| | Moderate slope | 0.00 | 0.00 |
| | Steep slopes | 0.00 | 0.00 |
| | Very steep slopes | 2.81 | 9.09 |
| | High plateau / mountainous | 0.00 | 0.00 |
| Annual crops (percentages): | Cereal | 21.13 | 28.57 |
| | Vegetables | 25.49 | 44.90 |
| | Leguminous | 1.74 | 2.04 |
| | Other cash | 43.57 | 67.35 |

not enough space left for fallowing, and often the yields are kept up only by the application of ever larger quantities of industrial fertilizer. In all likelihood under this system soil degradation due to lack of soil organic matter has already set in and will continue, so that ever more fertilizer will be needed, until the system will have to be abandoned.

On other sites especially in the more densely populated villages, the farmers apply manure on the cotton, which could be seen as a transitional step potentially able to avert organic matter depletion. At the same time, however, this step is not big enough, as the farmers themselves also assert, because there are not enough bush or other feed sources around any more to sustain the substantial number of cattle needed to produce enough manure. The rate of decline will therefore tend to be less steep than that of the fertilizer-based system, but with the same end result.

Theoretically, cotton could become sustainable if it were to become part of a highly integrated farming system where the on-farm production and recycling of all organic matter would get the utmost attention. This would amount to a further deepening and expansion of the indigenous intensification techniques mentioned in the previous section. The Mafa certainly have the knowledge and cultural capacity to make this possible, but it is hard to imagine that this could result in returns to labour (i.e. incomes) that could appeal to the new generation.

Maize is a crop that delivers more biomass than cotton and could therefore be easier to grow sustainably, by way of recycling the organic matter (through mulch, compost, manure etc.) *in situ*, supplemented by some nutrient subsidies from bushland or fertilizer. The same could hold for sorghum. It must be borne in mind, however, that farmers are aware of the ongoing soil degradation of the arable lands and yet do not apply these seemingly simple solutions.

The onions do not present sustainability problems (yet) and, as said, they could theoretically be a source for transition capital, through the direct and indirect (hired labour) incomes it supplies. The greater profitability per hectare of the larger-scale operations creates a tendency, however, that the scarce areas suitable for onions (vertisols close to the *mayos*) will end up in the hands of the lucky, powerful and urban few, who are disconnected from the rest of the plain and invest their profits elsewhere. The same story may hold true for the option of growing fruit trees such as mangos. As Timmermans and De Groot (2002) indicate, fruit trees are already spreading in north Cameroon, and market outlets could be improved. Moreover, the economy of scale may be different from the onions in that fruit trees may be more profitable for local growers than for urban-based outsiders. Fruit trees need good soils and irrigation, however, which implies that they tend to require the very same areas as do the onions, and spatial applicability will remain limited.

Seen in this light, the current trends of soil degradation, out-migration and poverty appear very difficult to reverse. Also, the spatial and economic context of the Koza does not appear to offer any of the external conditions that made the 'miracle of Machakos' (Tiffen et al. 1994) possible. There is no rapidly growing big city in the vicinity that could offer a market for new crops that could be grown on other soils than where the onions already are. Neither do the surroundings of the Koza plain offer large bush or vertisol areas that could act as external sources of organic matter. Fertilizer prices (coupled as they are with oil prices) are not likely to fall substantially, the climate is not predicted to improve and agricultural output prices are not predicted to rise substantially (other than during periods of drought that will hit the farmers just as hard as the urban people).

One possible scenario for a somewhat better future basically amounts to accepting a substantial out-migration and design a farming system that may use the lower population density, i.e. the larger space available per farm (say, 6 ha, at a population density of 100 people per km$^2$) to work against soil degradation and poverty. The crux of sustainability then would be to prevent that the larger average farm area would be used for just an expansion of the current unsustainable practices. Rather, a large part of the farm should be devoted to on-farm organic matter production, in order to keep an other part of the farm at a soil quality level comparable to or better than the current zones just around the farm houses (*champs de case*). Many variants of such a system are possible, working either with bush and fields separately or with smaller-scale mixtures of food and cash crops, forage and green manure crops, trees and grass, combined with soil and water conservation measures. The development such an 'integrated semi-intensive system' may be worked out in interactions between scientists, extentionists and farmers, taking into account that besides being the organic matter foundation for the farm, the bush or trees will also perform many of the other well-known bush functions, such as the procurement of firewood, medicines and fruits (Wassouni 2006). Intuitively, it would seem that the labour requirements of such a relatively relaxed system would be such that proper economic returns to labour may be reached.

In terms of the theoretical concepts, this would imply a divergence from the simple, dichotomous image that a viable mix of sustainability and acceptable returns to labour can be reached only in very extensive (pre-transitional) or very intensive (post-transitional) systems. This is theoretically interesting but may also serve as a warning against undue optimism about the intermediary system's viability. Would it really be technically possible, and profitable to an extent that farmers would be really motivated? How about the smallest farmers, whose areas may remain too small for this system to work? And how about the largest farmers, who may shift their cultivation yearly over rented plots and may be much more interested in access to (any) land than in land improvement?

In a wider context too, the simple notion of 'accepting the out-migration' carries a heavy practical and moral load. Where will these farmers go? Will they make a new living in the urban areas and possibly support the investments in their region of origin by way of remittances? The economic context of Cameroon does not give much hope for this scenario. Out-migrated farmers may well sink into urban poverty. Alternatively, farmers may migrate to recently opened-up areas further south, such as the Benoue region, but only to start a new cycle of slash-and-burn agriculture there, so that the whole process of degradation will repeat itself, only on a larger scale. Contextual questions such as these need to be addressed before a consistent long-term policy towards the Koza plain may be formulated, as is the case with the countless other places in the Sahel that are not endowed with good soil or nearby urban markets.

Chapter 8

# Forest Fringe Farmers on the Way to Sustainability: An Econometric and Cost–Benefit Analysis

Wouter T. de Groot and Marino R. Romero

Tropical forest degradation is commonly blamed on the slash-and-burn practices of upland farmers. These households are branded to be largely ignorant and unmotivated to implement soil and water conservation (SWC) techniques. In the Philippines, this has led to projects that confront farmers with pre-formatted farming systems that they are unwilling to adopt, often for good reasons, as was the case in Balete, which we describe later on.

An alternative way of looking that emerged recently recognizes that, partially or fully, many upland farmers do already transform their cultivation practices from slash and burn to more intensive and sustainable land use systems, e.g. through terracing, organic farming or agroforestry. Such scenarios are often called 'agricultural transition'. A better understanding of the factors that drive agricultural transition of forest fringe farmers will result in more effective policies and projects and may bring more farmers and more land into the transition process. Against this background, the objective of this chapter is to analyse the (proximate) factors that lead farmers in the Philippine uplands into agricultural transition.

## Theoretical Framework

Agricultural transition is linked to a number of explanatory perspectives on land use change. Some consider population growth as the (endogenous) driver of the transition process while others point at the (exogenous) market forces as the cause that motivates and capacitates farmers to make investments in sustainable land use.

The population paradigm consists of a pessimistic neo-Malthusian variant and more optimistic (neo-)Boserupian variants. From the Malthusian perspective, natural resource degradation is inevitable because of increasing population. A finite earth can only support a limited number of people. This theory is pessimistic about technological advances which, if within reach of people, may shift threshold levels and allow for an increase in food production.

The optimistic view on the effect of population on land use change is inspired by the seminal work of Boserup (1965). She points out that, facing land shortages, farmers will be inclined to invest in intensification even though, in the long run, this will tend to result in lower returns to labour. This process may in extreme cases ('involution', see Geertz 1963; Netting 2003; Hobbes 2005) lead to very high (and often sustainable) returns to land combined with very low returns to labour. Other population-centred authors, whom we will call 'neo-Boserupians' here, assert that in fortunate circumstances of soils and markets, the intensification may in fact lead to higher returns to labour. The description of Tiffen et al. (1994) of the 'miracle of Machakos' is a case in point. Conelly (1992) reports on a similar case in the Philippines, where irrigated rice and hillside fruit trees now provide higher incomes to more people than the original short-fallow swiddens.

As put forward by De Groot (1999b), cognitive and economic factors will co-determine which pathway will be taken by farmers or regions. When extensive farming methods become unsustainable under conditions of rising population density, some farmers may be aware early enough and have enough capacity to invest in the land and follow a neo-Boserupian road towards a new and sustainable system. Other farmers, however, may postpone the transition and enter a period of 'soil mining' e.g., because investments in soil and water conservation are less attractive than other options on the short term (Pender et al. 2004). These farmers may become motivated only at a time when they have no more capacity left, and are caught in a Malthusian poverty trap. Research of Murton (1999) has shown that even in the neo-Boserupian miracle region of Machakos, many farmers individually have gone the Malthusian way, ending up, for instance, as labourers making terraces on the very land that more successful, neo-Boserupian neighbours have bought from them. (Note that, in line with Platteau (2000), private land titles becoming prevalent in Machakos paved the way for this process of efficiency at the cost of equity.)

Writing on Uganda, Pender et al. (2004) show that many agricultural development pathways are market driven rather than population driven. Out of the group of more market-oriented and exogenous perspectives on agricultural transition we may take the neo-Thünian theory as explicated by De Groot (2006). In this perspective, large urban centres function as 'point markets' with zones around them (going from the city outward) of intensive agriculture, extensive agriculture and extraction of natural products. These zones are circular in a theoretically 'smooth' landscape and may be highly fragmented in practice. When 'point markets' grow, the zones that feed the markets begin to expand and as a result, farmers residing in a zone where only extensive agriculture was economically feasible before, may one day find that the economic 'intensification frontier' has passed their local area, along with the associated feeder roads, farm gate prices, extension, credit facilities, tenure security and so on. Thus the farmer will be inclined to switch to the now more attractive intensive options. Note that in this mechanism, local population density does not play any role.

As stressed by Pender (1998), Lipton (1989) and Netting (2003), the population- and market-based perspectives should not be applied as if mutually exclusive. The three population-based views differ only gradually, and the results of external market expansion intermingle freely with the effects of endogenous population growth. Each region will display its own mixture of mechanisms, and explanations, rather than work from one point of view, should focus on how this intermingling occurs and which of the mechanisms dominates – see for instance Zaal and Oostendorp (2002) and Chapter 2 of this volume discussing the case of Machakos in the light of both the population- and the market-driven points of view. Answers to these questions may also shift over time; a neo-Boserupian 'up' may be followed by a steady Boserupian decline; for instance, when the innovation potential cannot outstrip population growth any longer.

Against this background, this chapter focuses on the key element of agricultural transition: investment in the quality of the land (IQL). We take explanatory factors from both the market and population perspectives into account. The study sites are chosen from a basically Thünian perspective with varying distance from Manila, but local population densities are noted as well.

In order to act, people need to have both the capacity and the motivation to do so (Elster 1989; De Groot 1992; De Groot and Tadepally 2008). This simple truth is often overlooked in practice, e.g. in the Sustainable Livelihoods (SL) framework (www.ids.ac.uk/livelihoods) that puts all emphasis on 'capitals' and almost automatically leads to 'empowerment' policy recommendations, and in economic models where capacities only figure as 'constraints' and which tend to lead to motivational ('incentives') policy recommendations. More balanced explanatory models may of course be designed in many ways. Here, we will try our hand with a two-pronged approach. One element is an econometric analysis that, as we will see, puts most emphasis on capacity factors, and a cost–benefit analysis that has a fully motivational focus.

## The Study Sites

The data used in this chapter are mainly generated from a survey of 104 farmers living in four villages described below, which all display a significant presence of investments in the land and which are positioned along a gradient of distance to the markets of Manila – see Figure 8.1. Care was taken, moreover, to avoid correlation of distance to Manila and local population density, so as to be able to distinguish between the market and population effects in the later analysis. Although each village is not saliently different from others in its region, regional representativeness has not been a criterion and consequently we will not make any claims on the regional level. In each village, 26 household respondents were randomly selected, using a list of households kept by the *barangay* (village) secretaries. Jointly, the 104 households managed 235 plots.

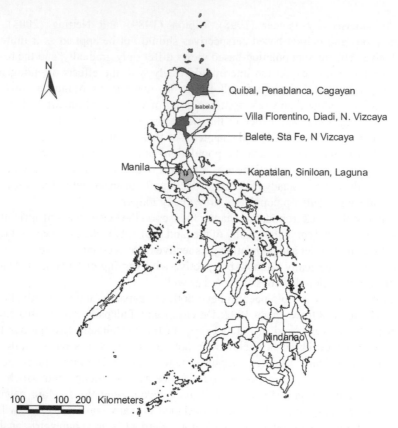

**Figure 8.1    Location of the study sites**

Kapatalan (population density 235 people per km²) is the most accessible among the *barangays*, connected as it is to Manila by a two-hour drive. Almost 90 per cent of the *barangay* area has slopes of 18 per cent and greater, with elevations ranging from 300 to 450 metres above sea level. From the time of settlement in the late 1950s, coconut and citrus have been the major agroforestry tree species grown. Under the coconut trees are papaya and root crops, namely gabi, taro and ginger. An Integrated Social Forestry Project was carried out in the village from 1988 onwards.

Balete (population density 13 people per km²) is a recently settled village where most of the households produce various kinds of vegetables, such as tomato, baguio beans, celery, carrots and string beans. For growing these vegetable crops, contour bunds are constructed by the farmers (to be remade every year), deviating from the hedgerow technology promoted by the Integrated Social Forestry Project that entered the village and declared it a model site in 1988. Most of the households are located about one to three kilometres away from the national highway that reaches Manila in some four hours.

At Villa Florentino (population density 38 people per km² and at nine hours' distance from Manila), the first wave of Ifugao migrants settled in 1973 followed by Igorot migrants the following year. The Ifugaos occupied the valley areas where they tapped water from creeks and later constructed rice terraces. The Igorots occupied the higher elevation parts of the village planting corn, upland rice and vegetable crops. Unlike the other villages, no government project was ever carried out here. Only the local government unit (LGU) of Diadi has interaction with the village officials.

Quibal (population density 93 people per km²) is located about 15 km from the urban market centre of Tuguegarao City, Cagayan, and some 13 hours from Manila. Large portions of Quibal lie within a declared Protected Area Landscape of the DENR (Department of the Environment and Natural Resources). Corn is the major crop grown in the village. Yellow corn varieties are sold on the feedstock market while the native varieties are planted for local consumption. Stimulated by a Community Forestry project that started in 1992, boundary planting of forest (Gmelina) and fruit (mango) trees species is the most common type of agroforestry adopted by the households. Fuel wood gathering provides a significant source of income.

**The Major Investments in the Quality of Land**

Table 8.1 presents the main characteristics of the major investments in land quality, grouped as terracing, contour bunds, irrigation facilities and tree planting. For ease of comparison with the labour requirements below, it is good to know that the average number of household members capable to work on-farm was about 3.52 and that accounting for age and school times, the total number of working days per year in a household was 766 man-days.

Terraces are established through the transformation of sloping land into productive areas where lowland rice and vegetables can grow. Terrace establishment requires high capital and labour inputs, ranging from 64 man-days per hectare in Quibal to as high as 1300 man-days in Villa Florentino. These variations reflect the material used in terracing as well as the slope of the land. This trend is similar in terms of cost (1998 prices, hired plus family labour) showing an average of Php 78,000 per hectare. Although costly, rice terraces allow two harvests per year of lowland rice, ensuring food security of the households as well as cash if surplus production is realized.

Forty-six plots in the villages were developed with contour bunds. The average productive area was about 0.49 ha while the total labour required per hectare of productive area was about 86 man-days, to be repeated every season because the bunds were only temporary constructions. The total cost per hectare (1998 prices, hired plus family labour) was about Php 8200 per year. Vegetables, considered to be high-valued crops, were grown in these areas, usually at two croppings per year.

**Table 8.1        The major investments in land quality adopted by households in Philippine villages**

| Investments in land quality | Means |
|---|---|
| *1. Terracing* | |
| Length, meters | 404 |
| Height, meters | 0.90 |
| Area, hectares | 0.40 |
| Labour required per hectare, man-days | 668 |
| Total establishment cost per hectare, Php | 77,905 |
| No. of plots with terraces in sample | 28 |
| *2. Contour bunds (yearly made)* | |
| Area, ha | 0.49 |
| Total labour required per hectare, man-days/y | 78 |
| Total cost per hectare, Php/y | 8,190 |
| No. of plots with contour bunds in sample | 46 |
| *3. Irrigation* | |
| 3.1 Channel irrigation | |
| Channel length, meter | 368 |
| Average labour required, man-day | 32 |
| Total establishment cost per meter, Php | 28 |
| No. of channels in sample | 26 |
| 3.2 Sprinkler irrigation | |
| Sprinkler length, meter | 751 |
| Average labour required, man-day | 4 |
| Total establishment cost per meter, Php | 11 |
| No. of sprinklers in sample | 47 |
| *4. Tree plantation (≥0.25 ha)* | |
| Area, ha | 1.26 |
| Average labour required per hectare, man-days | 27 |
| Total cost per hectare, Php | 2,300 |
| No. of plots with trees in sample | 88 |

Irrigation facilities are constructed to boost agricultural production in both rice terraces and contour bunds. Two types of irrigation facilities are used: the channel system and the sprinkler system. The channel system is mostly used for lowland rice cultivation while the sprinkler system is used in vegetable gardening. Households also used the sprinkler system for domestic water supply.

Tree planting activities were undertaken on 88 out of the 235 plots. In this study, 'tree planting' means that the area planted with trees is greater or equal to 0.25 hectare per plot. Households usually termed these as their 'agroforestry' or 'trees' farms; they required an average total labour of 24 man-days per hectare to establish them. This value is much lower than the other major investments.

**Results of the Econometric Analysis**

The econometric analysis of the factors influencing the investments is reported at length in Romero and De Groot (2008) and Romero (2006). The model utilized a logistic distribution which allows for the calculation of marginal effects of the explanatory variables. The dependent variables were major investments in land quality, aggregated into two categories. One category combines terracing, contour bunds and irrigation facilities, while the other category includes tree planting. These categories were selected considering the capital requirements and the time span of yields, as discussed above. The explanatory variables were grouped into two categories: household characteristics and farm characteristics. Table 8.2 gives the result of the regression analysis.

Table 8.2 shows significant correlations between the investments in the land and a number of explanatory household variables that will be briefly reviewed below, especially with respect to the question whether they represent capacity or motivational factors.

Age would be a motivational factor if the mechanism were that older farmers tend to invest more (as the sign of the correlation implies) because they desire to leave a sustainable and valuable farm to the children. Age would be a capacity factor if the mechanism were a life-cycle effect; older farmers have accumulated more physical, economic and social capital. The data do not allow a decision in this respect. Education, on the other hand, is probably a capacity factor; see also Pender and Kerr (1996). Off-farm and non-farming income, e.g. from remittances, trade or crafts, are often hypothesized as either a motivational or a capacity factor. On the one hand, they may demotivate farmers because they are less dependent on a sustainable farm (De los Angeles 1986). On the other hand, they may deliver farmers with extra income to convert into landesque capital (Reardon and Vosti 1995). If the sign is positive as is the case here, the capacity increase obviously dominates; the income is a capacity factor. Knowledge of soil and water conservation (SWC) techniques is a capacity factor.

A salient characteristic of Table 8.2 is the significant village dummies, about which more will be said later. These dummies capture a number of unmeasured village-level variables that may be either motivational (e.g. that the village needs good relations with environmental agencies or that the dominant ethnic group has strong traditions towards sustainable farming), or be capacity-related, e.g. collective social capital.

Overall, then, we may conclude that the econometric analysis has especially brought a number of capacity-related factors to the light. The major reason behind this is that the costs and benefits of the investments, which may be seen as the major motivational factor behind the investments, are largely determined by the

**Table 8.2　　The β-coefficients, odds ratios and probability values for the various investments in land quality[a]**

| | Investments in T, CB, and/or IF | | | Investments in AF and/or TG | | |
|---|---|---|---|---|---|---|
| | β | Odds ratio | Prob. value | β | Odds ratio | Prob. Value |
| *Household characteristics:* | | | | | | |
| Age | 0.06* | 1.06 | 0.07 | 0.09* | 1.10 | 0.04 |
| *Education: (w/ up to primary level as basis for comparison)* | | | | | | |
| 　Intermediate level | -0.52 | 0.60 | 0.53 | 2.92* | 18.59 | 0.02 |
| 　Secondary/College level | -0.88 | 0.41 | 0.31 | 0.59 | 1.80 | 0.63 |
| Household size | 0.15 | 1.16 | 0.51 | 0.03 | 1.03 | 0.95 |
| Off/non-farm and self-employment | 1.95* | 6.99 | 0.02 | -1.51 | 0.22 | 0.12 |
| Knowledge of SWC techniques | 0.36* | 1.43 | 0.07 | 0.20 | 1.23 | 0.37 |
| Security of tenure | 0.39 | 1.48 | 0.61 | 0.08 | 1.08 | 0.93 |
| Man-land ratio | 0.05 | 1.05 | 0.72 | -0.32 | 0.73 | 0.17 |
| With material asset | -1.34 | 0.26 | 0.30 | 2.55 | 12.83 | 0.15 |
| *Farm characteristics:* | | | | | | |
| Total landholdings | 0.10 | 1.11 | 0.43 | -0.12 | 0.89 | 0.46 |
| Ave. dist. to home | 0.0001 | 1.00 | 0.64 | 0.005 | 1.00 | 0.38 |
| Years of cont. cultivation | -0.01 | 1.00 | 0.30 | -0.18** | 0.83 | 0.01 |
| *Slope:(w/ steep slopes as basis for comparison)* | | | | | | |
| Flat (0-3%) | -1.93 | 0.14 | 0.18 | 0.61 | 1.84 | 0.73 |
| Rolling (4-8%) | -1.52 | 0.22 | 0.11 | 1.15 | 3.15 | 0.27 |
| Soil types: (w/ sandy loam as basis for comparison) | | | | | | |
| Clay loam | 0.25 | 1.28 | 0.73 | -0.06 | 0.94 | 0.94 |
| Village characteristics | | | | | | |
| *Village effect: (w/ V. Florentino as basis for comparison)* | | | | | | |
| 　Balete | 2.15* | 8.57 | 0.07 | 0.68 | 1.98 | 0.54 |
| 　Kapatalan | -1.31 | 0.27 | 0.44 | 3.80* | 44.87 | 0.09 |
| 　Quibal[b] | -2.06 | 0.13 | 0.24 | -3.19* | 0.04 | 0.07 |
| Pseudo R² | 46.94 | | | 60.48 | | |
| Number of observations | 95 | | | 95 | | |

*Note*: [a]T, CB, IF means that households invest in either terraces (T), contour bunds (CB) or irrigation facilities (IF); AF, TG means that households invest in either agroforestry (AF), or tree growing (TG).
[b]Missing value of the regression coefficient of this variable indicate non-variation of variable values at the household level.
* Indicates that the estimated coefficient is significantly different from zero at the 90 per cent level; ** indicates a significantly different from zero at the 95 per cent level.

prices of inputs (labour, seedlings, fertilizer etc.) and outputs (crops), which are invariable over the households and even the villages. With that, they are out of reach of the econometric (household-by-household) analysis.

## The Cost–Benefit Analysis

As said, the cost–benefit analysis (CBA) has sought to draw out the key motivational factors for investment in the quality of the land, complementary to the econometric analysis that appeared to shed light especially on capacity factors. The technical objection against CBA – that it tends only weakly to represent capacity constraints – does therefore not apply here, and for the same reason, we may focus on the internal rate of return (IRR) of the investments as the 'purest' motivational driver compared to the net present value (NPV) measure. We do add an indication of 'investment level' as a basic motivational factor to the overall schema however, in order to allow a comprehensive insight.

The CBA has an explanatory objective, i.e. seeking insight in why farmers apply IQL or not. This implies a focus on the household level rather than some wider system. Positive or negative externalities that are not expressed in prices or other motivational factors at the household level (e.g. through subsidies or levies) are therefore excluded from the analysis. Our 'emic' CBA is a financial one, focusing on cash flows at the farm gate. For the same explanatory reason, we take the household perspective as dominant. Sustainability, for instance, is important to farmers but largely in the form of expectations of soil quality rather than larger-scale issues of soil loss, biodiversity and the like. To take a final example, if the farmers do not know that international organizations are putting pressure on the Philippine government to terminate the price subsidy on rice, the relevant risk perception here is that the rice price is secure.

Our actor model is that of 'broad economic choice', which means that we refrain from cultural and social rationalities for land use (peer pressure, conviviality etc.), but assume that the actor is more than only a short-term profit (IRR) hunter. We have therefore taken up a number of factors that relate to risk and the relatively long term. One is the IRR that is realized not by the average price over the selling season but assuming that the farmer has to take the regularly occurring lowest (usually end-of-season) price. Rice is much more stable in this respect than tomatoes, for instance. The second factor is the degree to which farmers feel secure that prices may not fall dramatically in the somewhat longer run. For instance, will the fruit market be flooded by foreign produce, or will consumer taste turn away from what one is growing? A next factor is 'institutional backing', which stands for the degree to which farmers feel supported, e.g. by traders who supply credits for certain crops, by agricultural agencies which may back up the national staples with price guarantees or by environmental policies (e.g. those that supply de facto or *de jure* tenure security to farmers who are good for the land by making terraces or planting trees). Then there is the capacity of a crop to function as household staple, adding more to basic feelings of security. In our region, rice differs in this respect from all other crops (vegetables, corn, timber etc.) that need to be almost fully marketed. The final factor taken up is sustainability, standing, as said, for perceived soil quality stability under the crop. We have not tried to monetarize these factors but have taken them up as simple indicators (yes/no etc.) additionally to the two IRRs. As a whole, the data then come to look like an (emic) economic multi-criteria analysis with IRR central and other criteria added; see Table 8.3.

Family labour has been shadow-priced at the local market rates. All details of the CBA can be found in Romero (2006). The analysis is not differentiated between households, because the major factors determining the IRRs are much the same for all of them and because we already have the econometric data on inter-household differences. With that, the data requirements of the analysis are relatively low. The method has been nothing but a meticulous registration of all relevant prices and price variations, average investment cost (see Table 8.1) and running cost for the

**Table 8.3    Outcomes of the cost–benefit analysis and comparison with Bangladesh**

| Land use Option | Investm. level | IRR | IRR risk | L/term secure? | Instit. backing? | Auto-consum.? | Sustain-able? |
|---|---|---|---|---|---|---|---|
| Irr. Rice terrace | high | **40** | **30** | yes | yes | yes | yes |
| Irr. Veg. bunds | high | **400** | **90** | no? | yes | no | yes |
| Trees | low | **40** | **35** | ? | low | no | yes |
| Corn | zero | **50** | **-10** | yes | yes | no | no |
| Trees w/corn | low | **40** | **20** | yes | yes | no | yes? |
| Slash-and-burn | zero | **18** | **5** | yes | no | yes | no |
| | | | | | | | |
| BD annuals | zero | "40" | | yes | yes | yes | |
| BD agrofor. | high | "400" | | no | no | no | yes |

Core data are in bold.
IRR = internal rate of return (per cent per year), at average prices over the season.
IRR risk = internal rate of return with prices often occurring at the end of the season.
L/term secure = feeling of farmers that prices will not fall dramatically in the future.
Instit. backing = degree to which farmers feel supported by traders or government.
Auto-consum. = auto-consumability of the crop as household staple.
Sustainable = perceived level of long-term stability of soil quality under the crop.
Irr. rice terrace = gravity flow irrigated rice terrace.
Irr. veg. bunds = sprinkler irrigated vegetable on earth bunds.
Trees w/ corn = yellow corn with trees along plot boundaries.
Slashandburn = swidden agriculture on opened-up forest land.
BD annuals = rice or wheat in Bangladesh uplands (Rahman et al., 2008).
BD agrofor. = liche-based agroforestry in Bangladesh uplands (Rahman et al., 2008)
"40" and "400" = order-of-magnitude rates of return of the two systems in Bangladesh.

systems, average productivity and so on. This allowed a broadening of the scope of the analysis, and we have taken up two more land use systems in the data, namely cornfields without trees and slash-and-burn (swidden) agriculture. Yellow corn is the dominant land use type in the uplands. It is strongly dependent on fertilizers but farmers can get credits from corn traders that supply the feed markets for the piggeries around Manila. The major insecurity about this crop felt by farmers is that it exhausts the soils and requires ever more fertilizer to maintain yields, which often ends in farmers highly indebted with the traders (Hobbes and De Groot 2004). Slash-and-burn agriculture (called *kaingin* in the Philippines) is the major thing that upland farmers are supposed to do in many policy papers, associated with the dramatic loss of tropical rainforest that the Philippines have experienced. Farmers are often supposed to burn a new piece of forest once the fertility of the old piece has been exhausted, and knowing the economic profitability (IRR) of that action does of course help much explaining why they would do so, in lieu of staying put and invest in a settled land use system.

The CBA results can best be interpreted in light of the information that the local interest rate that farmers live with, i.e. the regular rate they can obtain from informal lenders, is approximately 15 per cent per year. This serves as a broad benchmark against which to compare the internal rate of return of land use. If a farmer works on credit, an IRR lower than 15 per cent means that he works at a loss and cannot fully repay the loan. If the farmer works without credit, doing something that generates less than 15 per cent implies that he could have earned more by lending the money out. Table 8.3 gives the result of the cost–benefit analysis.

With an internal rate of return at 40 per cent, irrigated rice terraces are a healthy if unspectacular investment especially because the risks are low. Price variations over the season can only reduce the IRR to 30 per cent. The price of rice is perceived as secure in the long run, as it is a national staple well backed up by market and government. This perception of low risk is reinforced by terraces being seen by the government as a good form of land use to be rewarded with secure tenure. Moreover, rice gives farmers certainty that they will always be able to feed their family and the terraces have a stable soil quality.

The vegetables on bunds show a very different picture. Profitability is very high at average prices (IRR = 400 per cent) but this tumbles to 90 per cent if only the seasonally lowest price is fetched. This uncertainty may well be felt concerning the longer term; what will happen if many other farmers enter the market?

Tree plantations are similar to rice terraces in terms of IRR and risk. The major difference is that the investment cost is much lower but, on the other hand, the farmer becomes more market dependent.

Growing corn without any investment in the land has been set here as requiring zero investment at the beginning of the season; this is only because traders supply credit, however. With an average IRR of 50 per cent, good profits can be made but the price variations and price sensitivity are such that the farmer runs at a significant loss if he is less lucky. Moreover, soil degradation results in that often, more and more fertilizer is needed and farmers run a severe risk that even the average profitability will plummet.

Planting trees along the cornfield boundaries slightly reduces the average rate of return but is much less vulnerable to the regular price fluctuations. Moreover, soil degradation will certainly be reduced thanks to the shade and leaf fall provided by the trees.

Slash-and-burn agriculture has a remarkable economic profile. It requires virtually no investment and it therefore accessible to the poor. The profitability is very low, however, compared to the other land use types and compared to the general local interest rate of 15 per cent, and the other economic factors do not add significant benefits to this picture. In other words, seeing it as we do here as a land use type of its own, slash-and-burn agriculture in this region is no economic undertaking at all.

Finally Table 8.3 shows the key characteristics of two farming systems in the uplands of Bangladesh, based on Rahman et al. (2008). One is wheat or rice in a simple monocrop system; the other is a liche-based agroforestry system using an economically clever timing for the intermediary crops in the first year, second year etc. Rahman et al. calculate NPVs rather than IRR but the overall picture is very clear: the NPV of the agroforestry system is ten times higher than of the monocrops. Assuming that the monocrop systems have a medium IRR comparable to the 40 per cent in our case, the '40' and '400' mentioned mention in Table 8.3 stand more or less symbolically to reflect the Bangldesh situation in our terms. We will use these data in the next section.

## Discussion

This section gives an overview and discussion of the substantive results. A broad methodological issue is picked up in the final section.

### *The position of slash-and-burn agriculture*

Slash-and-burn agriculture is, as we saw, not profitable in any direct sense in this region. Then why do people still do it? One explanation would be to point at the practically zero investment cost of this land use type. It could therefore be a 'refuge' land use for the poorest of the poor, who do not have access to anything else and also desire to be independent and stay outside the labour market, where casual labourers earn more than slash-and-burn farmers. This explanation, weak as it already is intuitively, is empirically untrue. As shown by Hobbes and De Groot (2006), these farmers are poor indeed in the sense that while carrying out the slash-and-burn agriculture they live a harsh life with little more than some rice on the table, in keeping with the very low IRR of this activity. But they do not stem from poorest of the poor backgrounds, and they know exactly what they are doing. What they are doing, as indicated by the CBA analysis and confirmed by the farmers themselves, is investing their labour in the establishment of a permanent (banana and corn) farm. New immigrants arriving with that desire find trees on the land they settle on. The trees have to go and to support the farmers during that process, farmers plant some upland rice. Slash-and-burn is nothing but the cutting edge of the expansion of permanent farming into the forest.

*Profitability is the great driver*

The cost–benefit analysis has an evident explanatory power for the overall changes in basic land use types. The step from an internal rate of return of virtually zero to one of some 40 per cent (which appears to be the basic figure in the region) is made spontaneously by the slash-and-burn farmers, as has been the establishment of vegetable farming, in spite of the high investments, in Balete, which offers even higher returns. In the area around 40 per cent – i.e. where farmers have to decide between maize, rice and trees – profitability does not play a dominant role any more. Farmers do still invest spontaneously in rice terraces (and sometimes in trees), but only if they have strong additional motivations (such as lower risk; see Table 8.3) and/or strong cultural inclinations as in Villa Florentino. Trees, for instance, bring hardly anything extra in terms of internal rate of return and, moreover, institutional support of trees through market parties and government is usually low, in spite of official government intentions (Masipiqueña et al. 2008). Usually, only strong projects can break through this lack of motivation to grow trees, e.g. supplying free seedlings and tenure security as extra motivators (Mangabat et al. forthcoming).

*The sum of additional factors is as strong as the immediate rate of return*

As we saw in the example of the trees, farmers may be 'tempted' into IQL even if the immediate rate of return does not improve significantly. The additional motivational factors (economic, social and cultural) are then brought into play. The strength of these factors is well illustrated by a case from the uplands of Bangladesh (Rahman et al. 2008) where, contrary to the Philippine data, one (agroforestry) land-use type had a spectacularly higher rate of return than annual cropping. Yet agroforestry was not adopted to any significant degree. Rahman et al. do not think that capacity factors play a dominant role in this; all farmers, for instance, know how or can learn to do agroforestry, and nothing prevents them from starting at a very small, home-garden scale. In the Bangladesh case, however, all additional factors were very strongly in favour of the annuals (see Table 8.3), e.g. through credit and price guarantees. As a summarizing rule, we may hypothesize that the sum of the additional motivational factors exerts an influence that is as strong as the immediate rate of return.

*Villages often unify farmers' responses*

The cost–benefit analysis has worked without internal variation within the region and can therefore not make visible, let alone explain, that farmers often act 'village-wise' (which is not to say that the villages are actors by themselves). With three significant village dummies, the regression analysis has shown, however, that the village level is a strong mediator in the farmers' land use decisions. It is unknown what factors play a role here. They may comprise, for instance, traditional knowledge and cultural preferences reinforced by social norms, collective social capital and leadership, copying behaviour starting out from one successful initiator, stimuli by traders, government and NGO projects, and physical/geographic circumstances (slopes, roads, soils, water), which are equal factors for most farmers in one village but may be quite different in the next.

*Do capacity factors play a role?*

If one declares not to be motivated for a good thing such as IQL, punishment or at least unpopularity may be expected. If one does not have the capacity to do the good thing, however, help may be expected. Capacity factors are therefore the favourites of many farmers (and of many development workers too, who may then go and do 'capacity building' and 'empowerment'). Do capacity factors really play a role in our case? Or are the IQL measures in fact so easy to spread out over the years and so easy to apply that any motivated farmer could find the capacity to do them, even if little by little? The regression analysis suggests that capacity factors do play a role to some extent. This seems clearly the case for the non-farm income that acts as a capacity boost for the (motivated) farmer. The factors of age and knowledge found in the regression analysis may be less unequivocal in this sense, but may also point at capacity constraints.

*More people, not more investment*

Returning briefly to the level of general land use theories, something may be said about the population-based (Malthusian, Boserupian, neo-Boserupian) and the geographic market-based (neo-Thünian) perspectives.

Population density may be measured with many scales, e.g. on the household, village or national levels. The man-to-land ratio is population density at the household level. Across all villages in the regression analysis, higher man-to-land ratios did not appear to coincide with more investments in the land. No relation with population is visible on the village level either. The village with the lowest population density (Balete), for instance, has a high investment level (see the village dummies), especially of the labour-intensive IQL types, and the most densely populated village (Kapatalan) was highest only in the least labour-intensive investment types. Quibal had the lowest level of investments overall, but ranked second in population density. Population-based theories of land use change do therefore not appear applicable to our case. (This finding does of course not rule out an effect of population density on the national level. That level is not what the population-based theories talk about, however, since the national population level works on the villages as an external force, e.g. through markets.)

*A Thünian pattern*

As put forward at the beginning of this chapter, a Thünian zoning of land use intensity around a large metropolis such as Manila can be expected to be fragmented due to all kinds of irregularities in the landscape. The vegetables in Balete, for instance, are there not only because of the relatively short distance of Manila but also because of the 'vertical landscape': the elevation and cool climate give Balete a comparative advantage over the low-lying plain around Manila. Temporal influences exert an influence, too. The short travel time from Kapatalan to Manila, for instance, is only recent; due to the new road, Kapatalan has experienced a 'zoning jump' from the extensive zone directly to a curious mixture of industrialized agriculture (poultry breeding) and other urban-based phenomena such as the construction of a resort and the buying up of forest patches by for recreational purposes, all of which are direct investments of Manila-based actors. In this perspective, it becomes more understandable that farmers in Kapatalan invest in trees rather than in the labour-intensive forms of IQL that are more characteristic of the intensive agriculture zone. The intensity of IQL of the other villages shows a more 'standard' Thünian pattern, as may be concluded from the village dummies of the regression analysis (Table 8.2). The village closest to Manila – Balete – is highest in both the labour-intensive and tree-based types of IQL. Next in terms of distance is Villa Florentino, with medium investments (the base village of the dummies). In the village farthest from Manila, Quibal, the level of both types of investments in the land were lowest.

## Metrics, Models and Transition Tendency Indicator: A Methodological Reflection

It has been only for intuitive reasons that an econometric analysis and a cost–benefit methodology have been applied here side by side to the same case; we wanted 'to have both' without exactly knowing why. In retrospect, it is only very seldom that two approaches are applied to the very same case, however, and this offers an opportunity for an 'interparadigmatic' methodological comparison.

Econometrics are inductive. Research starts out with only a weak statement about the world. Usually this is the formulation of a relatively large number of 'candidate factors' potentially explaining the phenomenon under study (in our case, IQL), and it is then left to the data to demonstrate which factors are important to which degree, set in some additive overall model; see Table 8.2. The CBA approach, on the other hand, is more deductive. It starts out with a strong statement about the world, proclaiming as it does that the CBA structure is the model of how farmers make decisions. The empirical work then is only to test or fill this model with the local facts; see Table 8.3.

Overmars et al. (2007) give a general discussion and illustration of inductive and deductive work in land use studies. They conclude that doing both is always better than following only one paradigm, and that land studies in general should try to become more deductive (theory driven). Something more than that may be said here, however.

Deductive work can be done with any number of research units, since models may be filled or tested with one, two or any number of households, villages or regions. Inductive work cannot be done with n=1, however. This, as we have seen here, results in a limitation of the inductive approach. Phenomena that are invariable over a region cannot be taken up in an inductive analysis of what actors in that region do. In our case, this implies that a full answer on the general question of *what drives IQL in this region* is out of reach of econometrics. Econometrics can only assess what explains inter-household (or inter-village) variation *within* the general drive or failure towards IQL.

The deductive CBA model has been applied here with n=1, namely with all farm-gate prices and other factors invariable over the whole region. This is no issue of principle, however. It would only require more data to apply the model at the village, distance-to-road, ethnic-group or household levels. Such a model could also include non-IRR and non-economic factors, as Overmars et al. (2007) show. In such an approach, econometrics would still play a supportive but important role, especially to get an empirically grounded grip on what factors to include in the model. In our case for instance, the age of household heads, non-farm income and some representation of village-level characteristics should be entered into the deductive household investment model.

What, in this context, would be a 'transition tendency indicator', i.e. some measure that could predict to what degree households would be inclined to invest in the quality of the land? Comparing the econometric Table 8.2 and the CBA model's Table 8.3, it is clear that any transition tendency indicator should be of the deductive, CBA-like type. The very low IRR of slash-and-burn agriculture versus the good IRRs of the other land use type is the indicator predicting that slash-and-burn is not a land use type of its own any more, and the small difference between corn and 'corn + trees' is the indicator predicting that communities hooked on corn will not spontaneously and *en masse* plant trees yet, while communities where the land is too steep anyway may switch from slash-and-burn pioneering to sustainable vegetable farming in large numbers and without government support.

A full transition tendency indicator model should of course be of greater subtlety than the rough CBA presented here. It should certainly incorporate capacity constraints and cultural inclinations, for instance, and specify how inductively gathered knowledge should be incorporated. Surely, however, it needs to be of the deductive models family and hence able to incorporate all relevant data, irrespective of whether they are variable or invariable across a region.

# Chapter 9

# Transition as Induced Innovation

Kees Burger

## Introduction

Fifty years ago, the semi-arid Machakos district in Kenya was a disaster area, characterized by overpopulation, soil erosion and poverty. Since that time the population has tripled, but so has per capita output, while soil erosion has virtually stopped (Tiffen et al. 1994). The 'miracle of Machakos' is a massive transition from unsustainable to sustainable agriculture, based on large-scale investment in terracing. Similarly Conelly (1992) describes how farmers in the Philippines, in an area at the fringe of the tropical rainforest, did not respond to the depletion of their soils in the usual way, by moving further into the forest, but by investing in irrigation and agroforestry, transforming their farms from shifting slash-and-burn to permanent sustainable systems. Such cases of agricultural transition through investments in sustainability have also been reported in Ghana, Nepal, Mali, Cameroon, and many other places (Fox 1993; De Groot and Kamminga 1995).

This chapter focuses on the relation between *transition* and *investment*. While these words were easily mentioned in the first paragraph above, their definition and their role in moving toward sustainable agriculture are not quite as clear. Investments imply the existence of some sort of capital, and the relationship suggests that a certain addition to this capital is needed in order to make the transition. As if 'capital' was lacking to make the transition before. We shall elaborate on the supposed relationship, which can be found in many analyses in the first section below. We then move on to set this investment approach against the background of the Boserupian transition to sustainable agriculture. One particular economic approach in this tradition is the *induced innovation* approach by Hayami and Ruttan, in which factor–price ratios induce research and innovation to adopt technologies that bring production techniques closer to the economic efficient combination of these factors. In the next section we show that the observed characteristics of the farms in Kenya and Benin confirm this idea of induced innovation, at least to the extent that the adoption of terracing and stone bunds led to labour intensities (man–land ratios) on the improved plots that are much higher than on not improved plots. The ratio of marginal products of land and labour (as a proxy for prices) also shows a change toward higher values. In the final section we conclude that the observed changes reflect a change in technologies, rather than investments, and that these technologies are adopted in response to changes in factor–price ratios, land scarcity in particular. In this perspective, the *transition* to sustainable agriculture is an adjustment toward changing economic conditions. Sustainability is a byproduct of the change, rather than the aimed-for outcome.

### The Investment Approach

While the word 'investment' sounds financial, we take the word in a wide sense, comprising the expenditure of labour, effort, time and agony and perhaps money to establish something durable. The element of durability is important in this approach. The basic idea is that agriculture itself erodes the durable asset, say soil fertility, and that the investment can help to restore this asset to level that can sustain high enough levels of production.

Households do not just autonomously take investment decisions. External effects, joint decisions, common properties, cultural aspects, and information all add a social dimension to the individual investments in sustainable farming and this must be taken into account.

The basic decision of the farmers is sketched in Figure 9.1. The farmer must consider two or more future paths to derive what economic returns and other benefits can be expected.

The term 'perceived' here indicates that the *subjective* evaluation by the farmers is the key issue. The term 'attainable' expresses that farmers must not only have the motivation to invest, but also the *capacity* (economic, intellectual and social capital; Reardon and Vosti 1995). Costs and benefits are here interpreted widely, including all characteristics that are normatively relevant to the farmer, e.g. the desirability of maintaining relationships in the community, and of staying close to the cultural image of what is a good farmer. 'Appropriateness' is the term given to these latter elements in De Groot (1992) and the benefits might better be coined *merits* to indicate the inclusiveness meant here. Some net merits can be measured directly as changes in income or expenditure, while others will have to

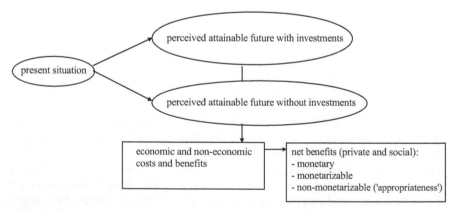

**Figure 9.1      Structure of the investment decision variables**

be evaluated in a model of expected utility maximization in which the household's valuation of various outcomes depends both on its risk aversion and on shadow prices (reflecting missing markets). Other merits may – by their nature – have to remain outside the model.

The question facing the farmer is then: Given a choice between a perceived future of continuing environmental degradation and a perceived future of environmental rehabilitation for which however investments have to be made, what do and will farm households choose?

The issue of rural investments in sustainability is linked to important theories. The 'Malthusian' paradigm, repeated in many variants in the environmental science literature, states that population growth irrevocably leads into a spiral of poverty and environmental degradation. Theory based on the work of Boserup (1965, 1981), however, claims that rising population densities lead to increased motivations and opportunities to innovate towards intensive and sustainable farming systems ('agricultural transition'). In fact, both phenomena can be observed. When, at rising population densities, fallows become too short to compensate for the nutrient loss during the cropping phase and environmental degradation sets in together with a decrease of returns to labour, there are two possibilities:

1. The 'poverty trap' scenario: by the time people are motivated enough to really start investing, they have become too poor to do so (e.g. Agbo et al. 1993).
2. The 'transition' scenario: people have become motivated while they are also able to invest, so that the 'Boserupian' path is taken.

Differentiation may take place within a single region and transition may be an option still open to the rich while it is no longer so for the poor.

Most of the investments will be of a physical nature (terracing, green manure, tree planting, livestock manure management etc.), but some may be non-physical too, e.g. investing in training, experiments or a village organization. Thus, many types of factor play a role and are addressed in the literature. To mention some from the literature, *soils* may be regenerable or not, or may be so acid that they force farmers into agro-forestry before they are caught in the poverty trap (Fujisaka and Wollenberg 1991). *Rural roads* reduce transaction cost and thus increase farm-gate prices, often for more sustainable crops; they also connect people with new ideas and extension agencies, raising their range of known land use options. *Remittances from male migration and off-farm income* may help the rural investors through the lean investment years; credit schemes partially work the same way. *Social capital and cultural inclinations* (leadership, exchange of knowledge, mutual help, attachment to the land (connected to time preference), gendered preferences of various options etc.) co-determine the motivations to invest. Also, farmers watch each other closely, and whole villages may either hesitate or suddenly go as one. In Machakos, male migration activated female social capital, which was then put to work in transition. *Market prices* for inputs and outputs obviously play a role in the trade-offs between the unsustainable and the alternative land use. Changes in prices may provide farmers the incentive to choose a Boserupian path but on the other hand, a subsidy on fertilizer may mask the need to do so (Ruben et al. 1996). *Tenure security* influences cultural attachment to the land as well as the economic considerations (Besley 1995). Or landlords may prohibit tree-planting options because it gives the planters rights to the land. *Perceived risks* may strongly influence motivations (even the risk of being gaoled for slash-and-burn activities, such as in Conelly 1992). The perceived attractiveness of *out-migration* may vary with preceived population densities in neighbouring rural areas and urban wages; see the 'gestion du terroir' literature, for example De Haan (1995).

This description of 'investment' and the capital that comes from cumulating investment, is couched in all-encompassing terms, and covers many aspects of human behaviour. Yet, underlying the approach is the acknowledgement of the existence of some capital, capital which apparently is needed to provide sufficient production and income and capital which yields returns in the near and more distant future. There is a time dimension involved, with present investments yielding future benefits.

In the economic analyses, this can be seen in a number of famous studies. Barbier (1996) used a dynamic optimization model in which future benefits were discounted to the time at which investments were made. Pagiola (1996) likewise uses a model in which future benefits are discounted (at 10 per cent). A typical result in his analysis is that investments in terracing initially reduce production and income, because of the land used for the terraces and the cost of constructing these, but later-on pay off, because the terracing stops the soil from degrading further. In a more recent review article, Knowler (2004) mentions 67 financial analyses from 10 studies in which net present value (NPV) calculations were done. Surprisingly, while about two thirds of the studies show positive NPVs for soil and water conservation (SWC) technologies in general, the return to investments in *structural* technologies such as terracing, is negative in the majority of the cases (14 out of 24).

## Technology Switch, Induced Innovation

Soil degradation is linked to the land use. At low levels of population density, people can feed themselves using the land extensively: after some years of use, other land can be taken into cultivation and the original area will have the time to recover naturally. This type of land use is still abundant in Africa. The mobility of the agricultural population itself is considerable and in many places land is abundant. Where labour is less mobile and the population growing, the demands on the land increase. The question is, then, what road will be followed?

The literature shows four approaches to this problem. The oldest one is from Thomas Malthus, who wrote in 1798 that food production would not be able to follow the growth of population, so that eventually population growth would be stopped by the availability of food. He, therefore, foresaw an equilibrium level for the population density at a low level of welfare, just enough to survive.

The second approach is from Esther Boserup who argued in her 1965 book *The Conditions of Agricultural Growth* that in times of increasing population density (and land scarcity) people shifted towards using technologies, often involving cattle, that made sustainable land use possible at higher levels of productivity. This made it possible to maintain the food production per capita. She describes this transition mostly as a social process in which the interaction between people is crucial for the realization of innovations. A more economic approach of the same transition is given by Hayami and Ruttan's (1981) *induced innovation* in which the change in technology depends on prevailing factor–price ratios.

The third approach is the neoclassical version of this process. The emphasis is on individual households for whom adoption of the new technology should be remunerative. Investments in terracing, for example, can become attractive when product prices increase faster than construction costs. Many recent studies try to show this by comparing benefits and costs. The importance of the approach is that they can show that many profitable investments are not made, simply because the money is lacking due to imperfections in the credit markets. The difference with the Hayami–Ruttam approach is the latter's emphasis on the price ratio of production factors and the innovation process, whereas the former gives individual profitability of adoption central stage.

The fourth and final approach builds upon the work of Von Thünen, who wrote in 1826 how the use of land is related to the distance to the market: more intensive near the market and more extensive farther away. Population growth in a region can lead to the formation of markets causing new outlets for agricultural products that may induce the use of other technologies and increase the value of the land.

In their analysis of the changes in Machakos, Tiffen et al. (1994) indicate that it were mostly Boserupian elements that played a role, while Nairobi's growing vicinity admittedly was an important factor as well, thus bringing in Von Thünen. They point at the growth of the population and increasing interaction among this population, an effect of more schooling and greater women involvement, to substantiate the Boserupian claim. The contribution by the nearer market was to facilitate migration and the transfer of remittances, but more importantly, to provide outlets for new and profitable products. In addition, the new crop, coffee in this case, provided a strong stimulus to create terraces on which trees could be planted.

As was mentioned in Chapter 1, the Tiffen and Mortimore group has continued its research along these lines. In Tiffen (2003), which refers to West Africa, more weight is given to the role of commercial opportunities than in the case of Machakos. They see provincial capitals as important engines of agricultural growth and the ensuing incentives towards sustainable management of the land. This regards land that can be reached from the centres. Improvements in infrastructure bring the centres closer to the surrounding land; thus, area that was in the outer circles of Von Thünen is brought within inner circles. The land correspondingly is now more suited for intensification, higher land prices result and profitable conservation can be undertaken.

The investment approaches by Barbier and Pagiola are in the neo-classical tradition, emphasizing farmer's rationality and focusing on the individual level. Hayami and Ruttan take a wider perspective, and account for changes in factor–price ratios. Their approach is shown in Figure 9.2.

Curve A–A is the set of possible combinations of land and labour that produce one unit (value) of output. It typically applies to a situation of abundance of land and is located in the south-west of the quadrant, while curve B–B shows the same unit isoquant for conditions where there is little land relative to labour. Both curves are valid for their sections in the quadrant: they show the immediately available technologies. When price ratios change drastically, gradually a new technology can be adopted, that is better adapted to the new price ratio. The hull of the local curves, here indicated by D–D, contains all the technical combinations of labour and land to produce one unit of output.

The transition from one technology to the other, and the investments in other forms of land are examples of *induced innovation* as described by Hayami and Ruttan (1985). If, in a situation with abundant land such as point P, land becomes scarcer, and therefore relatively more expensive, at first some adjustment might occur along the topical technology frontier, line A–A toward the north-west, with the optimal combination of labour and land shifting from a man–land ratio of ξ to one equal to θ, and from point P to point Q. At the same time pressure would be mounting to seek technologies that enable higher yields of the land so that less of

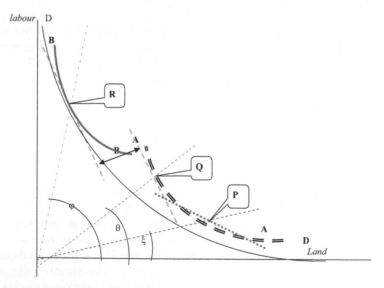

**Figure 9.2     Two technologies and *induced innovation***

this is needed to achieve the same unit of output: a shift toward line B–B and a man–land ratio of φ with lower costs, point R.

In their survey of effects of demographic changes on land quality, Templeton and Scherr (1997) sketch the same induced change but they use an extra step. They indicate that, yes, land intensity increases when relative factor prices change, but in their view the higher production per ha leads to soil depletion (lower soil fertility) and this is what encourages investments in SWC (p. 908).

It seems quite probable that the introduction of technology B–B requires more than just a modified factor land (such as terraced land), namely also other resources including fertilizer. Where access to such other inputs is restricted, the transition will not easily occur. Hence, market access plays a catalytic role. In a study of the relationship between poverty and soil degradation in various countries in Latin America, Swinton et al. (2003) reach the conclusion that the poor lack not just the means to invest in sustainable agriculture, but also the incentive. Investments would only help them if they had good access to the market. In the absence of such access, they continue working with the same technology and do not adjust the ratio of labour to land sufficiently.

Market access is not only important for access to inputs, it also provides an outlet for crops and closer markets signify higher prices for the crops. This increases the marginal products of labour and land in equal measure and would by itself not induce a change in technology. If the increased demand for labour can easily be met by increased supply of labour, the wages will hardly rise; the increased demand for land, however, may be less easily met by increased supply, and land prices are likely to go up. In this way, the higher crop prices, in combination with mobile labour, lead to similar incentives for a switch in technology as a direct change in the factor–price ratio.

Another effect of higher product prices is to overcome any transition costs that may exist. If the shift from technology A to technology B is not costless, a switch would only occur if the (future) benefits outweigh these costs. The change in operating costs can be seen from the double arrow showing the difference in the (dotted) iso–cost lines between Q and R in Figure 9.2. If it costs more per unit output than this to 'acquire' the technology, the change may not be made. Higher prices for the crop would increase the cost advantage of shifting to the new technology, so that the transition costs are more easily overcome.

The transition costs may take the shape of investments that are necessary to implement the new technology. This happens when terracing is required, and when this terracing constitutes a capital good. This capital-good nature rests on its capacity to provide services for a number of years, and at costs that are small relative to the establishment costs. If a terrace breaks down every year because of heavy rainfall, it is not a capital good. If maintenance costs are close to the establishment costs, it is hardly a capital good. We shall see that actually maintenance costs can be very high relative to establishment costs.

The investment approach to adoption of SWC is relevant where the devices used for SWC are costly capital goods. It should nevertheless also account for the change in technology and the more efficient use of scarce resources that the switch makes possible.

### Factors in the Machakos Case and Elsewhere

In the previous section, we indicated that a transition is studied toward a different technology that employs the factor land more intensively. We argue that this more intensive use of land responds to an increase in relative scarcity of that factor. In this section we address the underlying movement that led to scarcity of land in the first place.

*Increased population and scarcity of land*

As Tiffen et al. (1994) show, the population in Machakos dramatically increased over the decades between 1960 and 1990. On page 47 of the book, they indicate that population increased from 0.566 million in 1962 to 1.393 million in 1989, both figures measured within the 1979 boundaries. Thus at the end of the period, almost 2.5 times as many people were living in the same area. The area with population densities of more than 50 inhabitants/km² more than tripled over this period (Figure 4.3, p. 49). Much of this additional settlement occurred in agro-ecological zones of lower quality, partly by newcomers into Machakos, partly through out-migration from the densely populated areas.

To some extent the increased population settled in towns (about 0.1 million) but most of the additional population lived in the rural areas, and from the land that was cultivated where they lived. While in the early stages of the settlement hitherto unused land may have been available, toward the end of the era this was no longer the case. The relevant man–land ratio must have increased quite strongly therefore, making land relatively scarce.

This relative scarcity of land works as an increase of the factor–price ratio of land and labour. It calls for more labour intensive cultivation of the land, for example by choosing more labour intensive crops, and for soil conservation techniques that are labour intensive, such as terraces and stone bands. Following Eswaran and Kotwal (1985) more abundant labour also leads to lower wages for casual workers and fewer permanent contracts compared with seasonal work. Thus, crops that require peak labour, including coffee, become more attractive.

There are more effects of increased population density. As emphasized by Tiffen et al. (1994) the increased concentration of population improves the exchange of information, reduces the costs of assembling groups of people to work together, and improves the opportunities for local markets. In particular, people more easily learned about the technology of making the *fanya juu* terraces. In this sense, the population density improves the speed at which the farmers adopt terraces, once they consider them attractive enough.

Another, less frequently emphasized effect of population density is its propensity to lead to better infrastructure, such as roads, facilities such as water and sanitation, telephones and markets for visits by traders. While all this may have effects on production costs, they are more certain to positively affect the quality of life of the people. It is in their role as consumers that households appreciate such changes rather than in their role as producers. But once they enjoy these services, they feel themselves enjoying this particular location, and thereby enjoying the lie of their land. Eventually, it leads to a higher value of land in that place and this process therefore contributes to enforcing the price ratio of land and labour.

Finally, the improved infrastructure and the reduced costs of getting labour groups, and material needed for agriculture and terracing, reduces costs of farming and increases the farm-gate prices of the goods that are sold. The terms of trade of the farm households improves, in the same way as when the distance to the markets were reduced: purchased commodities become cheaper and selling commodities more rewarding. In particular, growing a cash crop for the market and using fertilizer becomes more attractive. To make efficient use of fertilizer, soil and water conservation measures are required, and these thus become more attractive.

If no SWC is applied, fertilizer use will be lower (because it would be less effective) and yields will be lower too. This may be a better choice for the farmer, if SWC is expensive and land cheap. Any effects of soil depletion or erosion can be avoided by shifting to a new parcel of land. This may explain why improving infrastructure may lead to more soil degradation as Pender et al. (2004) find for Uganda. If the improved roads, say, are not combined with scarcity of land, the incentive to invest in SWC is limited only to the direct yield-enhancing effects of retaining more water, seed and fertilizer in the field, but not in terms of value of the land.

*Technology change, investment or both?*

The more intensive use of land can take different forms. It can take the form of terraces, as in Machakos, but it can also be established by growing different crops, using different forms of crop rotation including reduced fallow, or adapting exchange patterns of land, as described by Mazzucato and Niemeyer (2000).

Whether the adoption of the new technology actually amounts to an *investment* depends on whether capital is required for this technology, in the sense of a durable production factor. If, say, terraces are important for application of the land-intensive technology and terraces do not wear out easily, then investments are required. If instead, stone bunds suffice to achieve the goal, and require restructuring every year, then there is no substantial investment involved, nor is 'capital' a production factor.

The distinction is important because when 'capital' is involved, this may pose a barrier for those that do not have sufficient resources to establish this capital. It also implies that one can 'add' to this capital or that it can erode. The benefits from a capital asset are spread out over time, so that in comparing costs and returns, one should take an inter-temporal framework with discounting of future benefits (and costs).

If, on the contrary, no capital is involved, the use of stone bunds or similar devices can be seen as part of the annual efforts that must be made toward harvesting a good crop. It poses no barriers to entry into the cultivation of the crop, and there is no immediate need for discounting any future benefits.

Many authors (Pagiola 1996; Barbier 1996; Hoogeveen and Oostendorp 2003; see Knowler 2004 for others) treat SWC, and terraces in particular, as capital goods, in which investments are made and that must be depreciated. In fact, they erode rather rapidly if not maintained. The approach implies that future benefits of the terrace are taking into account and that NPVs of future net benefits must be calculated for comparison with present costs of investment. It does require an estimate of the rate of decay of terraces, of maintenance costs and of establishment costs, and an assessment of a discount factor that the farmers have applied. Future returns are obviously uncertain, and their (discounted) value should account for this. The discount factor is typically high for poor farmers, and so is their degree of risk aversion. Hence, investment will only be undertaken if costs are very low (or expected returns very high). The usual approach is not to account for the uncertainty, and to take discount factors in the order of 10 per cent. Other evidence exists for much higher levels of discounting by poor farmers (Holden et al. 1998). If terraces can be established in lean periods in which the opportunity cost of labour is very low, the construction of such capital goods can be justified even with distant, uncertain and not-so-high returns. Another justification for the establishment of terraces may arise if the distinction between present and future is actually not so large, i.e. if the benefits of a terrace are enjoyed soon and if the difference between establishment and maintenance costs is small. Though the data are not very clear about this, there is in fact some evidence to substantiate this claim. The evidence in the data set for Machakos shows that the maintenance costs of a terrace are no less than about half the establishment costs (Table 9.1).

In addition, the establishment costs in terms of hours involved is not very high compared to other activities undertaken by the farmer, which lends some support to the assumption that terracing can be undertaken 'in spare time' and, as little cash costs are involved, also by poor people.

To be fair, there is other evidence too. Tiffen (1995, p. 200) indicates that maintenance costs are about 25 per cent of the establishment costs and the Philippine evidence of Romero shows maintenance costs of stone wall terraces to be 4 per cent of establishment costs and those of contour bunds 16 per cent (see Table 9.1). On the other hand, the annual costs involved in reconstructing the stone walls in the Atacora, Benin, or of anti-erosion devices in Cameroon are as high as the establishment costs, so that we should speak of an annual practice rather than an 'investment'.

If we take the stance of considering SWC as a technology choice, rather than as an asset, does this imply that there are no barriers to adopting the technology?

**Table 9.1    Labour (man days/ha) for construction and maintenance**

| (median values) | N | Man days |
|---|---|---|
| **Kenya** | | |
| construction of terrace | 96 | 12 |
| maintenance of terrace | 106 | 7 |
| **Philippines** | | |
| construction of stone terrace | 11 | 100 |
| maintenance of stone terrace | 12 | 4 |
| construction of contour bund | 44 | 85 |
| maintenance of contour bund | 5 | 14 |
| **Benin** | | |
| maintenance of stone bunds | 118 | 32 |
| **Cameroon** | | |
| construction of ingaleam | 35 | 3 |
| maintenance of ingaleam | 28 | 3 |

It does not. The new technology, even when it does not require a lump-sum investment, does require a different allocation and typically more of labour, and, in addition, it may require the use of other annual inputs such as fertilizer to be attractive enough. The data for Machakos and Benin show that on plots with some type of SWC higher yields are obtained, but also more labour is required. In Machakos, the production value per unit of labour input goes up, but in Benin, the extra labour required equals the higher yields so that production per man-day did not rise. And in Machakos, the higher production per day also serves to pay for additional inputs such as fertilizer.

To the extent that access to such external input is restricted, the farmers have limited incentives to switch to another technology. These restrictions can be external (no trader visits the village) or internal (household does not have sufficient funds to advance the money for fertilizer). In both cases, the household will find itself restricted in its choice, and will choose the likely sub-optimal option of non-adoption.

In addition to fertilizer, the farmers should have knowledge of how to farm the plot with SWC. This may be trivial, but it may also imply the cultivation of a new crop, coffee in Machakos for example, for which new expertise must be gained. As long as farmers lack this knowledge, they may choose not to adopt the technology.

Such restrictions on their freedom pose barriers to adopting the technology, similar to the assumed asset nature of SWC: if one does not have enough information, money or assets, one does not adopt; once overcome, the barriers cease to exist and the farmers with sufficient knowledge or cash to pre-finance fertilizer or asset, will see no restrictions in applying the new technology.

If farmers face a barrier, their observed position as to adoption of SWC could be suboptimal and any estimated marginal effect of SWC need not be equal to the marginal costs of establishing the SWC. The gradual adoption of terraces in Machakos and Kitui, as shown by Zaal and Oostendorp (2002), reflects both the changes in attractiveness over time, as prices of outputs, inputs and production factors changed, and the changes in the 'capital' of the farmers, consisting of their access to outside inputs, knowledge of the technique and knowledge of the technology of growing more rewarding crops in the terraced fields. As more farmers gained sufficient knowledge, more adopted SWC. Thus the adoption curve is the result of many influences, both in the region and in the farm households themselves.

This process of accumulating knowledge of techniques that are adequate for the new situation of more labour per hectare is what is behind the shift in the isoquants of Figure 9.2 from point Q to point R. The known combinations of land and labour that produce one unit of output changed from those represented by the curve A–A gradually to those represented by curve B–B. Upon reaching point R no known technologies exist that would reduce the production costs of unit of output any further.

### Conclusions

Adoption of terracing helps reduce erosion, and thereby contributes to more sustainable farming. This has led a number of authors to investigate if this was indeed the motive for farmers to adopt terracing. Did they see their land deteriorating and did they foresee that the process needed to be stopped by investing in soil and water conservation devices? Was it the deterioration of land that motivated the investment? We tried to argue in this chapter, that this was not the case. Land need not become less productive to induce farmers to invest in SWC. Land only needs to become more valuable. A foremost condition for this to happen is for land supply to be limited. Once this is the case, there are various ways in which land becomes more valuable. One way is when better prices are paid for the products, for example because towns are more easily reached. Another way is if increasing population demands more land. This is a particularly strong force as it also leads to relatively cheap labour, making a labour intensive technology more attractive. A third way could be deterioration of land. While this lowers the value of the average plot, it makes fertile land scarcer. We argue that this deterioration is by no means needed for a transition toward a new, land saving technology. The mere change in the relative price of land *vis-à-vis* labour would justify such a change. In the case of Machakos we see both the first and the second process occur.

We also argued that the change does not require a capital investment. The work to be done for SWC can in general be done with the work force of even poor households. The maintenance costs of SWC are high compared with the establishment costs. Other factors are at least as important for the adoption of a new technology as SWC investment, including access to inputs and knowledge of how to make the transition and how to apply the new technology. Thus the change toward farming on terraced fields is more the result of the attractiveness of the new technology in view of the change in relative factor prices than it is a forward-looking response to impending loss of soil fertility.

# Chapter 10

# Lessons on Land Management

Fred Zaal and Kees Burger

## Introduction

In the tradition of every discipline or study there are books that mark a transition from one paradigm of thinking to the next, though there are questions as to whether these shifts exist in social science (Nederveen Pieterse 1998). The study of human population growth and the limits of its environment, or the extent to which this population can provide itself with adequate food, energy and raw materials in a limited area, certainly provides examples of such books. These books, while marking the transition, provide inspiration and guidance for newly emerging paradigms.

One of the earliest contributions to the discussion on population growth and living standards is the study by Malthus (1798). His model implied that populations were continuously at the margin of survival. A growing population would experience decreasing returns to labour in relation to a limiting factor of production. In Malthus' model, the limiting factor of production was land. With decreasing production per capita, that population would be at equilibrium with potential production on that limited land. Whenever technologies would allow an increase in productivity, population growth would increase until the new equilibrium was achieved, expanding the population, not improving its standard of living in the long run. In view of the long history of very slow to almost absent growth of the world population as a whole for most of its history (despite occasional rapid growth rates in particular areas in very limited periods), this model was accurate and valid. As Galor and Weil (1999) indicate, basing their work on Maddison (1982), for most of Europe's history during the Middle Ages, population growth followed productivity growth and GDP per capita was virtually stable. Technology change led to higher density populations rather than higher standards of living, so that England's living standards at the time of Malthus' writing were roughly equal to those of ancient Roman or Han Chinese living standards (Easterly 2003). And population growth between the years 1 and 1750 was well below 1 per cent per year (Livi-Bacci 1997). However, ironically, at around the time of Malthus' writing, this pattern started to change, with population growth slowing down, at least in Western Europe, while productivity growth increased. Though fertility increased at first and mortality declined, the rapid technological change allowed an even more rapid increase in productivity and GDP per capita, and a slow increase in the standard of living until the mid-nineteenth century, when the pace of improvement accelerated. After that, population growth slowed down.

## Innovation, the Optimistic View

Within this context, a second book has tended to influence the discussion on whether it is possible to have a rapidly growing population while maintaining productivity growth and improving the standard of living, through technological development. The study of Esther Boserup (1965) implied that there is a two-way causality between population growth and the development of technologies. The slow pace of agricultural innovation in Africa was ascribed to lower initial levels of population pressure, and the absence of markets for which higher productivity-level technologies could be employed. In such areas and certainly after the introduction of colonial rule, labour would leave the area rather than be employed to improve productivity, thus extending the phase of relatively low population pressures (Boserup 1985). Under these conditions, it will take longer for certain technologies to be adopted, even when they are well known to the land users. They simply are not the appropriate technologies for the conditions at hand. Her study focused attention on the balance between population pressure, labour availability, land availability, market development and technological development being achieved in an intricate and causal deterministic and thus optimistic scenario in relation to the number of people and their income per head, but also in terms of sustainable agriculture. However, this process is not always unproblematic. It may well be that limits are encountered in relation to land or other factors of production. If these causal interactions are indeed relatively deterministic, we can expect the process to take place everywhere, and at all levels of scale. This is not what we see in practice, where long periods and large areas seem to be outside this process of induced innovation. Around the early 1980s, despair about the future of rural populations in many developing regions was getting the better of development researchers (e.g. Goldschmidt 1981).

A third book, by Tiffen, Mortimore and Gichuki (1994) tried to establish this pattern of increasing population pressure, increasing adoption of first inappropriate and later more appropriate technologies, that would allow a continuous increase of population, productivity and higher standards of living in a concrete setting: Machakos District, Kenya, and for a long period of time: the last sixty years. The book and the associated journal articles were welcomed as a positive contribution to social environmental science after two decades of growing pessimism about the world's environment, and the environment in Africa in particular. The world needed 'success stories'.

The publications also came in handy at a time of privatization and neoliberalism, as it did not put emphasis on government-driven 'development', but on people's own capabilities. In the Machakos case, population growth and more competition for scarce resources did not necessarily lead to doom and disaster, to conflict and war, but stimulated innovations, agricultural intensification, and more environmental care. The authors dropped the diagnosis–prescription–intervention model that had so dominated the history of dryland management in Machakos (and most of Africa, for that matter) since 1930, and instead focused on the resourcefulness of society. In the period 1930–90 a trend of degradation was turned towards sustainability in resource management; while the population grew at world-record speed: a five- or even six-fold increase during those years. The long view – 60 years – directs analytical attention away from the intervention–response framework to a confrontation with the processes of economic and environmental change (Tiffen et al. 1994, p. 16). The result is the 'alternative nexus of positive interactions' (ibid., p. 261).

Amidst the general praise, one of the criticisms on the book was the lack of household-level data that could be used to develop a model and test the thesis of its causal relationship more rigorously. Certainly, these trends of population growth and environmental recovery had coincided, but what was the relationship between these trends? What made farmers effectuate the recovery, and were other factors than population growth not as important? These were the central questions that led us to the start of our research programme.

### Discounted Value of Preventing Future Erosion

An important consideration at the start of the research programme was the nature of the anti-erosion device. It was typically considered a capital good, for which a substantial investment was required and which would last some years with minor maintenance costs. The rock and stone terraces surely fit this description, but stone bunds, contour bunds and many other devices are often almost completely reconstructed every year. And some terraces, notably those made of earth and grass, require maintenance costs that are not small compared with the establishment costs.

The physical characteristics of the devices are, however, not the only reason to consider soil and water conservation (SWC) a capital good. The issue of sustainability requires a long-term perspective, and the call for change is based on the allegedly negative long-run consequences of present behaviour. The soil quality itself is therefore the capital good, which in many studies is believed to be eroded year after year, and this erosion is to be stopped by switching to the new technology. Thus, the studies by Pagiola (1996), Pender and Kerr (1998) and others include among the long-term benefits of the investment in terraces the non-deterioration of the soil. Default is a long-term downward trend, and the positive scenario is where this trend is reversed.

A problem with the long-term models is its obligatory discounting of future benefits, with the discount rate set at some rather arbitrary value. Other studies that focused on how poor farmers discount the future often show outcomes for the discount rate that far exceed the values included in the long-term models. Yesuf (2004) comes at 43 per cent for Ethiopian farmers, while Holden et al. (1998)

report values for farmers in Zambia, Ethiopia and Indonesia that are even higher. Indeed, if these high discount values prevailed, hardly any investment would be economically justified. This is the more so as returns (in terms of erosion prevented) come only gradually over time, first small and eventually large, but then also heavily discounted.

### Market Access for Some …

Why, then, would farmers still switch to a soil conservation technology, if it were not for conserving soil fertility? One answer is that returns can be reaped quickly if the technology enables the use of fertilizer and/or the growing of profitable annual crops. Another answer is that the costs of the investment can be very low if only slack-season labour input is required. The first answer requires access to the market, for inputs as well as for the output. Chapters 2 and 4 focused on this aspect. The second answer (cheap labour) is more applicable to poor and remote farmers who have no alternative use for family labour outside the peak seasonal activities (as in Benin, see Chapter 7). In both cases any returns to the switch can be reaped fairly quickly. If, as is the case in some parts of Kenya, coffee trees are planted on the newly terraced plots, a longer term perspective *is* relevant. These returns come only in future years. With high discount rates, such investment in terracing and tree planting is only undertaken if expected returns are very high indeed. In all cases however, returns, measured in terms of future soil fertility maintained, may play a very modest if not negligible role.

The above considerations explain why distance to the market matters so much. When returns to SWC depend so much on access to fertilizer and the market for output, its use is likely restricted to those farmers that can access the market. More distant farmers will not invest, not even when the costs are low, as there is no pay-off. In the course of economic development, more roads and better communication may connect farmers better to the markets, thus contributing to better returns to improvements in land quality. The Von Thünen reasoning applies here: farmers closer to the centre show higher land intensities. With growing cities, the 'circle' of such high land intensities will broaden and encompass more and more farm households.

The most distant farm households are those that are effectively isolated from the city. The distance to this centre then loses its relevance, and the group of farm households actually may become their own centre. When land is scarce, as was the case in the insecure mountain regions of north Cameroon (see Chapter 7), households invest in SWC to produce as reliably as possible the much-needed food from the scarce resources. Their innovations are not focused on productivity growth, but on stability of yield, and they end up choosing for the same technology.

## ... Land Scarcity for Others

Households that have invested in SWC show man–land ratios on these plots with SWC that are considerably higher than those on plots without SWC. Such higher labour intensity on SWC plots would make sense if land was or had become scarcer relative to labour, as Burger argues in Chapter 9 of this volume. Land becomes more valuable relative to labour when more workers enter a region with a given amount of cultivable land. Such was the case in Machakos, but only after some time. Taking more land into cultivation could accommodate the initial inflow of people. Yet, this brought them to – at least initially – less attractive fields, with less access to markets, fewer amenities etc. and probably smaller farms. To make a living on smaller farm sizes with a certain family endowment of labour, SWC would help in creating employment for all this labour. These households therefore adopted SWC and a technology justifying the higher labour intensity. Other farms however were not affected by this scarcity of land. The old existing farms may not participate so much in the land market. They may have sold some land, or rented out some plots, but it is likely that most would simply stick to their land. Their man–land ratio did not change, unless their families grew or more hired labour was employed, as more workers came to seek employment in Machakos. But not so many households did hire in workers, according to our data, and most of them worked with their own family labour or organized labour parties with the farmers in their neighbourhood. These groups mostly operate outside of and are insulated from the labour market. With no transactions in either the land market or the labour market, households could remain untouched by the changes in the relative scarcity of land, and hence would not change their technology (as long as output prices do not change). Land scarcity therefore, though it should lead to higher (rental) prices for land, does not affect all households, and many may remain untouched. This is in fact what we see in the case of Kenya (see Chapter 3), where the decision to introduce terracing and other improvements such as the application of manure and fertilizer is made predominantly in the first few years after a plot has changed hands, both through inheritance and through the selling and purchasing of land. This phenomenon links the regional drivers of change to dynamic processes at the level of the household, farm and individual. This is unlike the case for, say, a change in the price of maize, which would affect all households other than the pure subsistence households. Changes in the price of labour are also likely to affect more households than changes in land scarcity, as households more often hired out or hire in labour than land.

## Von Thünen

It was von Thünen who asserted that, under certain conditions, a pattern of decreasing intensity of land use with distance would have to appear around the major markets for which these areas produce and from which they obtain their inputs. One condition, a homogenous plain with transaction costs reduced by distance from the market, is never met in practice of course, and so this pattern interferes with local agro-ecological conditions and opportunities, and the layout and quality of infrastructure. Our findings related to the distance to the market (at the regional level) of Nairobi or Manila for example, and the transaction costs (at the level of the households and plots within the farm) support this model. In fact, at the level of the region, the reduced distance to market (in terms of travel time) is the predominant factor in explaining adoption of SWC over time. Transaction costs as a factor, and its interaction with other variables, play an important role in the explanation of investments taking place at the level of the farm and the plot. This supports the view that it is economic – market – opportunities rather than other factors that drive the adoption of innovations such as terracing.

The actual layout of infrastructural improvements and the actual characteristics of the terrain are not the only diversions between the concentric model of von Thünen and the real world. The question of population densities in the area surrounding the markets is also a matter of concrete study. There is of course an interaction between population densities and the layout and quality of the infrastructure, as the one will follow the other in many instances. In our research, we have controlled for this and identified the impact of each factor. The influence of market access came out as much stronger than that of population density. The weak impact of the latter can be explained by its flexibility, for out-migration as a process may influence this interaction. It reduces population density while at the same time making capital more abundant through remittances. In areas with less labour but more access to capital, a shift to techniques may occur that makes the best of these trends, such as permanent structures with lower maintenance.

## Level

Of the factors that we discussed, some are at the level of the plot (its inherent fertility, slope), some at the level of the household (position in its life cycle, man–land ratio, wealth), some are at the level of the village (distance to major market, population density and population pressure) and some at higher levels such as prevailing prices for crops, infrastructure and security. These levels were distinguished to answer the question of which factors are responsible for a massive change toward sustainable agriculture. Of the factors at the regional and village level that Tiffen et al. (1994) considered, our analysis on the basis of farm household data shows that distance to the market is the strongest factor. The closer the market and the more involved households are in market transactions, the more investments in SWC. This was particularly true for Kenya (Machakos and Kitui) and for the Philippines. In the more remote regions of Boukombé in Benin and the Koza plains in Cameroon such distance factors are less important. Only a weak effect of distance on the adoption of SWC was found for the Beninese farmers. When access to market is important, this implies that output prices and the availability of inputs and their prices must be important. Thus high coffee prices helped Machakos farmers switch to terracing, and good returns to vegetables induced terracing in the Philippines. These are influences that in principle affect all households in the village the same way. All can in principle start growing these crops. Some may be more remote than others despite being in the same village, but by and large all could make the transition. Adoption of such new crops requires flexibility and the drive to undertake this and this may favour younger and more educated household heads over older ones. When, however, sufficient land suitable for growing such crops is available, the incentive to adopt SWC diminishes. Scarcity of such land helps strengthen the incentive to convert existing sloped land into land suitable for cultivation of these promising crops.

Chapter 5 by Gandonou and Oostendorp showed that reliable evidence on the beneficial effects of SWC could be found only when controlling for (observable and unobservable) differences across households. Apparently, fertility and treatment of the plots differ so much between households, that effects of SWC are blurred by these idiosyncrasies at the household level. This suggests that the adoption of SWC is hampered by market imperfections, making it rational for some households to adopt and for others to continue without SWC investment. With improving market access and/or increasing land scarcity, more and more farmers can be expected to make SWC investments, but this remains a dynamic process with some households leading and some households following. Chapter 3 by Oostendorp and Zaal shows that within this dynamic process the (intergenerational) transfer of land forms an important trigger as well, as a large part of the SWC investments is typically taking place within a few years after the land is transferred to a new owner.

**Other Factors**

Of the other factors that authors on this topic found to be relevant we discuss two important variables that contribute to the switch to SWC, namely the size of the household and the farm, and the household's degree of off-farm employment. Farm size and household size appeared to matter in the same sense, i.e. that the relevant variable actually is the man–land ratio: household size divided by farm size. The higher the man–land ratio, the stronger the incentive to switch to a labour-intensive technology, so that the negative effect of household size can only be understood when farm size is even higher for larger households. This appears to be the case. In Kenya, apparently farm size by itself has a negative effect of the probability of (having invested in) SWC, while household size by itself has little effect.

Off-farm income intervenes in this relationship in many ways. High levels of off-farm income may imply that more household members are living on the farm but not *off* the farm. It might also induce the household to focus on off-farm opportunities rather than on-farm investments. On the other hand, off-farm income, for example remittances, may provide the household with the money to hire working groups to establish terracing and to make the further outlays to engage in cash crop production (planting material, fertilizer). The empirical analysis of this type of relationships is hampered by the cross-section nature of the data that were collected. Much terracing was established some time ago, and benefits from growing crops in these fields may have led the household to adjust its household size, farm size, and its range of activities. Hence the currently observed non-farm income may be the result of earlier terracing decisions rather than its explanation. Panel data are required to investigate how terracing contributes to further changes in the livelihood strategies of the household and perhaps more specialization in farming.

**Towards a Transition Indicator Model?**

As mentioned in the preface, the initial aim of the research programme was to derive a sort of 'transition tendency indicator model' which could predict the probability of a region making a transition toward sustainable land use. The problematic aspect of this question is that it presumed the original situation to be not sustainable. Such perceived non-sustainability typically occurs in situations where land is either abundantly available, or where the population has little interest in the future value of the land (because their use of the land is temporary) or in the rare event that population has the incentive but not the means to move to sustainable land use as would happen when large investments must be made, or when these require coordination problems to be solved. The cases of Machakos and Kitui show that the alleged non-sustainability of land use was probably solved when land became sufficiently scarce, and demand for its products sufficiently strong. With respect to north Cameroon, Chapter 7 shows that the pacification of the Koza plains opened up large tracts of new land for the terrace-builders of the Mandara mountains; with land scarcity no longer an issue, SWC was dropped. The Benin case in the Atacora showed how remote farmers typically invested in SWC to sustain their livelihoods, and might even invest a little more when they would be connected to the market. Migration is likely to attenuate further intensification, however.

Thus, fewer people and more land apparently induce *less* SWC. The converse, more people and less land, led, in the Machakos case, to *more* SWC. Would therefore a measure of population density suffice to indicate transition to sustainable farming? Obviously not (as emphasized by Dietz and his co-authors in Chapter 1), as density does not imply pressure on the use of the land. To have density translate into pressure, agriculture should be the dominant economic activity. This can be so if agriculture is most rewarding or, equivalently, other activities are less remunerative. And within agriculture, the attractiveness of crops that make intensive use of soil and water conservation is especially important.

Any indicator of the likelihood that SWC will be adopted should reflect such attractiveness. What costs and benefits, and what time perspective should be taken into account here? In more cases than we initially expected, SWC appears an annual undertaking. Only for the robust stone terraces (important in the Philippines) the capital-good nature of the SWC and thereby the possible constraints on the capacity to make the investment are important. In other cases the change in technology has more to do with *motivational* aspects than with *capacity* aspects. As De Groot and Romero show in Chapter 8, a straightforward application of benefit–cost analysis focusing on recurrent costs comes a long way in characterizing the differential attractiveness of a crop with and without SWC. This measure of immediate returns does not reflect the aspect of land scarcity. As Chapters 9 and 3 show, scarcity is important for adoption, yet, scarcity is not felt by each and every farmer: it is mostly the new farmers, or expanding farmers (and other participants in land transactions) that notice the dearness of the land and take this into account when deciding on SWC. A regional aspect that promotes adoption is therefore scarcity, combined with dynamics in land-ownership.

In conclusion, we find that remunerative agriculture favours adoption of sustainable technologies; and we find that increased scarcity of land, relative to labour, helps moving in this direction. Neither of the two forces is by itself sufficient, however. Where remunerative agriculture coincides with land abundance, sustainability is threatened, however, and the combination of land scarcity with non-remunerative agriculture leads to poverty, auguring migration out of agriculture.

But combined, a remunerative agriculture and relative land scarcity creates the conditions for sustainable land use. This happened in Machakos. It may occur in many other places.

# References

Ade Freeman, H. and J. Smith (1996), 'Intensification of Land Use and the Evolution of Agricultural Systems in the West-African Northern Guinea Savanna', *Quarterly Journal of International Agriculture* 35 (2), pp. 109–24.

Adégbidi, A. (2001), 'Etude des Filières des Intrants Agricoles au Bénin', rapport de consultation pour l'IFS (MDR/GTZ) (Cotonou).

Adégbidi, A., E. Gandonou, E. Padonou, H. Océni, R. Maliki, M. Mègnanglo and D. Konnon (2001), *Etude des filières des intrants agricoles au Bénin: Rapport de consultation pour l'IFS* (Cotonou: Ministère de Développement Rural/ Deutsche Gesellschaft für Technische Zusammenarbeit).

Adégbidi, A., E. Gandonou and R. Oostendorp (2004), 'Measuring the Productivity from Indigenous Soil and Water Conservation Technologies with Household Fixed Effects: A Case Study of Hilly-Mountainous Areas of Bénin', *Economic Development and Cultural Change* 52, pp. 313–46.

Admassie, Y., L.I. Mwarasombo and P. Mbogo (1998), *The Sustainability of the Catchment Approach-induced Measures and Activities* (Nairobi: National Soil and Water Conservation Programme, Ministry of Agriculture).

Agarwall, A. (2001), 'Common Property Institutions and Sustainable Governance of Resources', *World Development* 29 (10), pp. 1649–72.

Agbo, V., N. Sokpon, J. Hough and P.C. West (1993), 'Population-Environment Dynamics in a Constrained Ecosystem in Northern Benin', in G.D. Ness, W.D. Drake and S.R. Brechin (eds), *Population-Environment Dynamics* (Ann Arbor: University of Michigan Press, pp. 283–300).

Alexandratos, N. (1999), 'World Food and Agriculture: Outlook for the Medium and Longer Term', *Proceedings National Academy of Sciences* 96 (May), pp. 5908–14.

Allan, W. (1965), *The African Husbandman* (Edinburgh: Oliver and Boyd).

Anderson, J.R. (1980), 'Nature and Significance of Risk in the Exploitation of New Technology', in J.G. Ryan and H.L. Thompson (eds), *Socio-Economic Constraints to Development of Semi-Arid Tropical Agriculture* (ICRISAT, Hyderabad, India), pp. 297–302.

Anderson, J.R., J.L. Dillon and J.B. Hardaker (1977), *Decision Analysis in Agricultural Development* (Ames: Iowa State University Press).

André, C. and J.P. Platteau (1998), 'Land Relations under Unbearable Stress: Rwanda Caught in the Malthusian Trap', *Journal of Economic Behavior and Organisation* 34 (1), pp. 1–47.

Antle, J.M. (1987), 'Econometric Estimation of Producers' Risk Attitudes', *American Journal of Agricultural Economics*, pp. 509–522.

Antle, J.M. (1989), 'Nonstructural Risk Attitude Estimation', *American Journal of Agricultural Economics* 71, pp. 774–783.

Antle, J.M. and C.C. Crissman (1990) *Risk, Efficiency, and the Adoption of Modern Crop Varieties: Evidence from the Philippines – Economic Development and Cultural Change* (Chicago: Chicago Press).

Argwings-Kodhek, G., M. Mutua, and G. Okumu (1990), 'PAM Analysis of Agriculture in Kisii and Siaya Districts. Policy Analysis for Rural Development', Working Paper No. 5 (Njoro: Egerton University).

Azontondé, A., I. Youssouf and A. Akakpo (1995), *Etudes Morphopédologiques des Bassins Versants constituant le Site de Boukombé* (Cotonou: Centre National d'Agro-Pédologie (CENAP) et Projet de Gestion des Ressources Naturelles (PGRN)).

Babatoundé, L. and A. Sounkoua (1996), *Diagnostic Foncier sur le Site de Boukombé* (Cotonou : Projet de Gestion des Ressources Naturelles (PGRN)).

Baland, J.M. and J-Ph. Platteau (1996), Halting Degradation of Natural Resources: Is there a Role for Rural Communities? (London: Clarendon Press).

Barbier, B. (1996), 'Impact of Market and Population Pressure on Production, Incomes and Natural Resources in the Dryland Savannas of West-Africa: Bioeconomic Modelling at the Village Level', EPTD Discussion Paper no. 22 (Washington DC; IFPRI).

Barbier, B. and G. Bergeron (1998), *Natural Resource Management in the Hillsides of Honduras: Bio-economic Modelling at the Micro-Watershed Level*, Environment and Production Technology Division (Washington DC: IFPRI).

Bardhan, P. (1984), *Land, Labour and Rural Poverty* (New York: Columbia University Press).

Besley, T. (1995), 'Property Rights and Investment Incentives: Theory and Evidence from Ghana', *Journal of Political Economy* 105 (5), pp. 903–37.

Bevan, D., P. Collier and J-W. Gunning (1992), 'Anatomy of a Temporary Trade shock: the Kenyan Coffee Boom of 1976–79'. *Journal of African Economies* 1(2), pp. 271–305.

Bhattacharyaa, A. and S. Kumbhakar (1997), 'Market Imperfections in the Presence of Expenditure Constraint: A Generalized Shadow Price Approach', *American Journal of Agricultural Economics*, 79 (August), pp. 860–71.

Bigman, D. (1996), 'Safety-First Criteria and Their Measures of Risk', *American Journal of Agricultural Economics*, 78, pp. 225–35.

Bigsten, A. (1996), 'The Circular Migration of Smallholders in Kenya', *Journal of African Economies* 5 (1), pp. 1–20.

Billon P. le (2001), 'The Political Ecology of War: Natural Resources and Armed Conflicts', *Political Geography* 20 (5), 561–84.

Binns, T. (ed.) (1996), *People and Environment in Africa* (Chichester: Wiley and Sons).

Binswanger, H.P. (1982), 'Econometric Estimation and the Use of Risk Preference', *American Journal of Agricultural Economics* 64, pp. 391–3.

Born, G.J. van den, M. Schaeffer and R. Leemans (1999), Climate Scenarios for Semi-arid and Sub-humid Regions: a Comparison of Climate Scenarios for the Dryland Regions in West Africa from 1990 to 2050, ICCD Report 2 (Wageningen/Bilthoven: AUW/RIVM).

Boserup, E. (1965), *The Conditions of Agricultural Growth: the Economics of Agrarian Change Under Population* (New York: Aldine).

Boserup, E. (1981), Population and Technological Change: A Study in Long-term Trends (Chicago, IL: University of Chicago Press).

Boserup, E. (1985), 'Economic and Demographic Interrelationships in Sub-Saharan Africa', *Population and Development Review* 11 (3), pp. 383–7.

Boulet, J. (1971), Magoumaz: étude d'un terroir de montagne en pays Mafa (Yaoundé : ORSTOM).

Bowen, M. (2005), 'The Impact of Risk on Land Investments in Semi-Arid Lands of Kenya', PhD thesis (Moi University, Eldoret, Kenya).

Boyd, C. and T. Slaymaker (2000), 'Re-examining the "More People Less Erosion" Hypothesis: Special Case or Wider Trend?' *ODI Natural Resource Perspectives* 63 (November) (London: ODI).

Brons, J., F. Zaal, R. Ruben and L. Kersbergen (2000), *Portfolio Diversification and Rural Development Pathways: A Village Level Analysis in South Mali*, ICCD report (Wageningen/Amsterdam: AUW/UvA).

Brown, L. and R. Kane (1994), *Full House. Reassessing the Earth's Population Carrying Capacity*. The World Watch Environmental Alert Series (New York: Norton).

Brown, S. and B. Shrestha (2000), 'Market-driven Land-use Dynamics in the Middle Mountains of Nepal', *Journal of Environmental Management* 59 (3), pp. 217–25.

Buch-Hansen, M. (1992), Towards Sustainable Development and Utilisation of Natural Resources in Arid and Semi-arid Rural Areas of Africa, Research Report No. 85 (Roskilde University).

Bunch, R. and G. Lopez (1995), *Soil Recuperation in Central America: Sustaining Innovation after Intervention* (London: International Institute for Environment and Development).

Burton, B., D. Rigby and T. Young (2003), 'Modelling the Adoption of Organic Horticultural Technology in the UK using Duration Analysis', *Australian Journal of Agriculture and Resource Economics* 47 (1), pp. 29–54.

Byiringiro, F. and T. Reardon (1996), 'Farm Productivity in Rwanda: Effects of Farm Size, Erosion and Soil Conservation Investments', *Agricultural Economics* 15, pp. 127–36.

Cameron, A. and P. Trivedi (2005), *Microeconometrics. Methods and Applications* (Cambridge, UK: Cambridge University Press).

Campbell, D. (1981), Soils, Water Resources and Land Use in the Mandara Mountains of North Cameroon, MSU/USAID Mandara Mountains Research Report No. 14.

CARDER (1969–1999), annual reports (Natitingou, Atacora).

Chamberlain, G. (1980), 'Analysis of Covariance with Qualitative Data', *Review of Economic Studies* 47, pp. 225–46.

Chayanov, A.V. (1966), *The Theory of Peasant Economy*, eds D. Thorner, B. Kerblay and R.E.F. Smith (Illinois: Homewood).

Chopra, E. (1989), 'Land Degradation: Dimensions and Casualties', *Indian Journal of Agricultural Economics* 44 (1), pp. 45–54.

Christensen, L.R., D.W. Jorgenson and L.J. Lau (1971), 'Conjugate Duality and the Transcendental Logarithmic Production Function', *Econometrica* 39, pp. 255–6.

Churchill, N.C. and K.J.Hatten (1997), 'Non-market Based Transfers of Wealth and Power: a Research Framework for Family Business', *Family Business Review* 10 (1), pp. 53–67.

Clay, D. and T. Reardon (1994), 'Determinants of Farm-level Conservation Investments in Rwanda', Occasional Paper No. 7, International Association of Agricultural Economists (IAAE).

Clay, D.C., F. Byiringiro, J. Kangasniemi, T. Reardon, B. Sibomona, and L. Uwamariya (1995), *Promoting Food Security in Rwanda through Sustainable Agricultural Productivity: Meeting the Challenges of Population Pressure, Land Degradation, and Poverty*. Staff Paper No. 95-08 (March) (East Lansing, MI: Department of Agricultural Economics, Michigan State University).

Clay, D., T. Reardon and J. Kangasniemi (1996), 'Sustainable Intensification in the Highland Tropics: Rwandan Farmers' Investments in Land Conservation and Soil Fertility', *Economic Development and Cultural Change* 46 (2), pp. 353–77.

Conelly, W.T. (1992), 'Agricultural Intensification in a Philippine Frontier Community: Impact on Labour Efficiency and Farm Diversity', *Human Ecology* 20 (2), 203–23.

Conelly, W.Th. and M.S. Chaiken (2000), 'Intensive Farming, Agro-diversity and Food Security under Conditions of Extreme Population Pressure in Western Kenya', *Human Ecology* 28 (1), pp. 19–51.

Coombs, C., R. Dawes and A. Tversky (1970), *Mathematical Psychology* (Englewood Cliffs NJ: Prentice Hall).

Costin, A.B and H.C. Coombs (1981), 'Farm Planting for Resource Conservation', *Search* 12 (12), pp. 429–30.

Coxhead, I. and S. Jayasuriya (1995), 'Trade and Tax Policy Reform and the Environment: The Economics of Soil Erosion in Developing Countries', *American Journal of Agricultural Economics*, 77 (3), pp. 631–44.

Crowley, E.L. and S.E. Carter (2000), 'Agrarian Change and the Changing Relationships between Toil and Soil in Maragoli, Western Kenya (1900–2000)', *Human Ecology* 28 (3), pp. 383–414.

D'Emden, F., R. Llewellyn and M. Burton (2006), 'Adoption of Conservation Tillage in Australian Cropping Regions: An Application of Duration Analysis', *Technological Forecasting and Social Change* 73, pp. 630–47.

De Groot, W.T. (1992), Environmental Science Theory: Concepts and Methods in a Problem-oriented, One-world Paradigm (Amsterdam: Elsevier Science Publishers).

De Groot, W.T. (1999a), *A Future for the Mountains: A Policy-oriented Synthesis of Three Recent Dissertations on the Mandara Mountains, North Cameroon* (NIRP Programme/CML Leiden University, Leiden).

De Groot, W.T. (1999b), *Van vriend naar vijand naar verslagene en verder; een evolutionair perspectief op de relatie tussen mens en natuur* (Nijmegen: Nijmegen University Press).

De Groot, W.T. (2006), 'From Friend to Enemy and Onwards: An Evolutionary Perspective on the People–nature Relationship', in R.J.G. van den Born, R.H.J. Lenders and W.T. De Groot (eds) *Visions of Nature* (Berlin: Lit Verlag).

De Groot, W.T. and E.M. Kamminga (1995), *Forest, People, Government* (Leiden: Leiden University).

De Groot, W.T. and H. Tadepally (2008), 'Community Action for Environmental Restoration: a Case Study on Collective Social Capital in India', *Environment, Development and Sustainability* 10 (4), pp. 519–536.

De Haan, L. (1995), 'Vers une utilisation durable de l'environnement dans le departement du Borgou (Benin)', in P. Ton and L. de Haan (eds) *A la recherche de l'agriculture durable au Bénin*, Amsterdamse Geografische Studies 49, pp. 121–6.

De la Briere, B. (1999), 'Determinants of Sustainable Soil Conservation Practices' Adoption: An Analysis for the Dominican Republic Highlands', paper presented at the Economic Policy Reforms and Sustainable Land Use in LDCs: Recent Advances in Quantitative Analysis (Wageningen, 30 June–2 July 1999).

De los Angeles, M. (1986), Economic Analysis of Households in Upland Community (Makati, Manila: PIDS).

Dietz, T. (1987), *Pastoralists in Dire Straits. Survival Strategies and External Interventions in a Semi-arid Region on the Kenya/Uganda Border: Western Pokot, 1900–1986* (University of Amsterdam).

Dietz, T. and M. Put (1999), 'Climate Change in Dryland West Africa and the Resultant Response of the Local Population, their Institutions and Organisations', in F. Von Benda Beckmann and O. Hospes (eds) Proceedings of the Ceres Seminar 'Acts of Man and Nature: Different Constructions of Social and Natural Resource Dynamics', Bergen, 22–24 October 1998 (Wageningen: Ceres).

Drylands Research (2001), 'Livelihood Transformations in Semi-arid Africa, 1960–2000', proceedings of a workshop arranged by the ODI with Drylands Research and the ESRC, in the series *Transformations in African Agriculture*. Drylands Research Working Paper 40.

Dzuda, L.N. (2001), 'Analysis of Soil and Water Conservation Techniques in Zimbabwe: a Duration Analysis', MSc thesis (Agricultural Economics, Dept of Rural Economy, University of Alberta).

Earle, T.R., C.W. Rose and A.A. Brownlea (1979), 'Socio-economic Predictors of Intention Towards Soil Conservation and Their Implication in Environmental Management', *Journal of Environmental Management* 9 (3): 225–36.

Easterly, W. (2003), *The Elusive Quest for Growth: Economists' Adventures and Misadventures in the Tropics* (Cambridge: MIT Press).

Ellis, F. (1998), 'Household Strategies and Rural Development Diversification', *Journal of Development Studies* 35 (1), 1–38.

Ellis, F. (2000), Rural Livelihoods and Diversity in Developing Countries (Oxford: Oxford University Press).

Elster, J. (1989), *Nuts and Bolts for the Social Sciences* (Cambridge: Cambridge University Press).

Ersado, L., G. Amacher and J. Alwang (2004), 'Productivity and Land Enhancing Technologies in Northern Ethiopia: Health, Public Investments, and Sequential Adoption', *American Journal of Agricultural Economics* 86 (2), pp. 321–31.

Ervin, C.A. and D.E. Ervin (1982), 'Factors Affecting the Use of Soil Conservation Practices: Hypothesis, Evidence and Policy Implications', *Land Economics* 58 (3), pp. 277–92.

Ervin, D.E. (1986), 'Constraints to Practicing Soil Conservation: Land Tenure Relationships', in S.B. Lovejoy and T.L. Nappier (eds), *Conserving Soil: Insights from Socio-economic Research* (Ankeny, Iowa: Soil Conservation Society of America).

Eswaran, M. and A. Kotwal (1985), 'A Theory of Two-tiered Labor Markets in Agrarian Economies', *American Economic Review*, March, pp. 162–77.

Fafchamps, M. (1992a), 'Cash Crop Production, Food Price Volatility, and Rural Market Integration in the Third World', *American Journal of Agricultural Economics* 74 (1), pp. 90–99.

Fafchamps, M. (1992b), 'Solidarity Networks in Pre-Industrial Societies: Rational Peasants with a Moral Economy', *Economic Development and Cultural Change* 41 (1), pp. 147–74.

Fafchamps, M. and F. Shilpi (2003), 'The Spatial Division of Labour in Nepal', *Journal of Development Studies* 39 (6), pp. 23–66.

Fall, A. (2000), 'Makueni District Profile: Livestock Management', Drylands Research Working Paper 8 in the Kenya Series of Policy Requirements for Farmer Investment in Semi-Arid Africa (Somerset: Drylands Research).

Fan, S. and P. Hazell (1999), 'Are Returns to Public Investment Lower in Less-favored Rural Areas? An Empirical Analysis of India', EPTD Discussion Paper No. 43 (Washington DC: IFPRI).

Fauck, R. and R. Maignien (1959), 'Mission d'études au Dahomey', rapport de pédologie No. 2 (Sols de Boukombé: ORSTOM).

Faye, A., A. Fall, M. Tiffen, M. Mortimore and J. Nelson (2001), 'Région de Diourbel: synthesis', Drylands Research Working Paper 23 (Somerset: Drylands Research).

Feder, G., R.E. Just and D. Zilberman (1985), 'Adoption of Agricultural Innovations in Developing Countries: A Survey', *Economic Development and Cultural Change* 33, pp. 255–98.

Feder, G.R. and O'Mara G.T. (1981), 'Farm Size and Adoption of Green Revolution Technology', *Economic Development and Cultural Change* 30, pp. 59–76.

Flavian, K. and D.A. Hoekstra (1990), 'Dryland Management: The Machakos District, Kenya', in *Dryland Management: Economic Case Studies*, edited by John A.D.

Fleuret, R. and A. Fleuret (1991), 'Social Organisation, Resource Management and Child Nutrition in the Taita Hills, Kenya', *The American Anthropology* 93, 91–114.

Fox, J. (1993), 'Forest Resources in a Nepali Village in 1980 and 1990: The Positive Influence of Population Growth', *Mountain Research and Development* 13 (1), pp. 89–98.

Fujisaka, S. and E. Wollenberg (1991), 'From Forest to Agroforest and from Logger to Agroforester: A Case Study', *Agroforestry Systems* 14 (2), pp. 113–29.

Galor, O. and D.N. Weil (1999), 'From Malthusian Stagnation to Modern Growth', *The American Economic Review* 89 (2), pp. 150–4.

Gandonou, E.A. (2006), 'Sustainable Land Use and Distance to the Market – Micro Evidence from Northern Benin', PhD dissertation (VU University, Amsterdam).

Gandonou, E. and A. Adégbidi (2000), 'Croissance démographique et dynamique des systèmes paysans dans les régions Montagneuses du Bénin: L'intensification agricole, en panne?' (Université Nationale du Bénin).

Geertz, C. (1963), *Agricultural Involution* (Berkeley CA: University of California Press).

Geist, H.J. and E.F. Lambin (2001), 'What drives Tropical Deforestation?' Global Environmental Change-Human and Policy Dimensions 11, pp. 261–9.

Geotechnip (1969), Etude de Reconnaissance des Sols. Région de Boukombé (République du Dahomey), Seine et Oise (France), et Cotonou: Geotechnip et Ministère du Développement Rural (Cotonou: Institut National de la Statistique et de l'Analyse Economique).

Ghodake, R.D. and J.G. Ryan (1981), 'Human Labor Availability and Employment in Semi-Arid Tropical India', *Indian Journal of Agricultural Economics* 36 (4), pp. 31–8.

Gichuki, F.N. (2000a), 'Makueni District Profile: Farm Development 1946–1999', Drylands Research Working Paper 1 in the Kenya Series of Policy Requirements for Farmer Investment in Semi-Arid Africa (Somerset: Drylands Research).

Gichuki, F.N. (2000b), 'Makueni District Profile: Rainfall Variability, 1950–1997', Drylands Research Working Paper 2 in the Kenya Series of Policy Requirements for Farmer Investment in Semi-Arid Africa (Somerset: Drylands Research).

Gichuki, F.N. (2000c), 'Makueni District Profile: Water Management, 1989–1998', Drylands Research Working Paper 3 in the Kenya Series of Policy

Requirements for Farmer Investment in Semi-Arid Africa (Somerset: Drylands Research).

Gichuki, F.N. (2000d), 'Makueni District Profile: Soil Management and Conservation, 1989–1998', Drylands Research Working Paper 4 in the Kenya Series of Policy Requirements for Farmer Investment in Semi-Arid Africa (Somerset: Drylands Research).

Gichuki, F.N. (2000e), 'Makueni District Profile: Tree Management, 1989–1998', Drylands Research Working Paper 5 in the Kenya Series of Policy Requirements for Farmer Investment in Semi-Arid Africa (Somerset: Drylands Research).

Gichuki, F., S. Mbogoh, M. Tiffen and M. Mortimore (2001), 'Makueni District Profile (2001), Synthesis', Drylands Research Working Paper 11 in the Kenya Series of Policy Requirements for Farmer Investment in Semi-Arid Africa (Somerset: Drylands Research).

Gleave, M.B. and H.P. White (1969), 'Population Density in Agricultural Systems in West Africa', in M.F. Thomas and G.W. Whittington (eds), *Environment and Land-use in Africa* (London: Methuen).

Goldschmidt, W. (1981), 'The Failure of Pastoral Economic Development Programs in Africa', in J.G. Galaty, D. Aronson, P.C. Salzman and A. Chouinard (eds) *The Future of Pastoral Peoples* (Ottawa: IDRC).

Gould, B. (1996), book review of 'People and Environment in Africa', by T. Binns, *Transactions of the Institute of British Geographers* 21 (2), pp. 436–7.

Gould, B.W., W.E. Saupe and R.M. Klemme (1989), 'Conservation Tillage: The Role of Operator Characteristics and the Perception of Soil Erosion', *Land Economics* 65, pp. 167–82.

Grabowski, R. (1990), 'Agriculture, Mechanization and Land Tenure', *Journal of Development Studies* 27 (1), pp. 43–53.

Gray, L.C. and M. Kevane (2001), 'Evolving Tenure Rights and Agricultural Intensification in Southwest Burkina Faso', *World Development* 29 (4), pp. 573–87.

Greene, W. (2000), *Econometric Analysis*, 4th ed. (London: Prentice Hall International).

Haegerstrand, T. (1967), *Innovation Diffusion as a Spatial Process* (Chicago, IL: University of Chicago Press).

Hall, R.E. (1973), 'Wages, Income and Hours of Work in the U.S. Labour Force', in G.G. Cain and H.W. Watts (eds) *IncomeMaintenance and Labour Supply* (Chicago, IL: Markham).

Hardaker, J.B., R.B.M. Huirne and J.R. Anderson (1977), *Coping With Risk in Agriculture*, (Wollingford: CAB International).

Havens, A.E. (1975), 'Diffusion of New Seed Varieties and its Consequences: a Colombian case', in R.E. Dumett and L.J Brainard (eds) *Problems of Rural Development: Case Studies and Multidisciplinary Perspectives* (Leiden: E.J. Brill).

Hayami, Y. and V. Ruttan (1985), *Agricultural Development. An International Perspective* (Baltimore: Johns Hopkins University Press).

Hayes, J., M. Roth and L. Zepeda (1997), 'Tenure Security, Investment and Productivity in Gambian Agriculture: a General Probit Analysis', *American Journal of Agricultural Economics* 79 (May), pp. 369–82.

Heckman, J. and T. MaCurdy, T. (1986), 'Labor Econometrics', in Z. Griliches and M.D. Intriligator (eds) *Handbook of Econometrics* (Amsterdam: Elsevier).

Heyer, J. and J. Waweru (1976), 'The Development of Small Farm Area', in Heyer, J., J. Maitha and W. Senga (eds) *Agricultural Development in Kenya: An Economic Assessment* (Nairobi: Oxford University Press).

Hobbes, M. (2005), 'Material Flow Accounting of Rural Communities: Principles and Outcomes in South East Asia', *International Journal of Global Environmental Issues* 5 (3/4), pp. 194–222.

Hobbes, M. and W. T. de Groot (2004), 'Slopes, Markets and Patrons: Socio-economics along a Lowland–upland Gradient in the Philippines', *Pilipinas* 43, 71–91.

Holden, S.T. (1991), 'Peasants and Sustainable Development – the Chitemene Region of Zambia. Theory, Evidence and Models', PhD dissertation (Ås: Agricultural University of Norway, Department of Economics and Social Sciences)..

Holden, S.T., B. Shiferaw and W. Mik (1998), 'Poverty, Market Imperfections and Time Preferences: of Relevance for Environmental Policy?' *Environment and Development Economics* 3, pp. 105–30.

Holloway, G., Nicholson, C., Delgado, C. Staal, S., Ehui, S. (2000), 'Agro-industrialization through Institutional Innovation: Transaction Costs, Cooperatives and Milk-market Development in the East-African Highlands', *Agricultural Economics* 23 (3), pp. 279–88.

Hoogeveen, H. and R. Oostendorp (2003), 'On the Use of Cost/Benefit Analysis for the Evaluation of Farm Household Investments in Natural Resource Conservation', *Environment and Development Economics* 8, pp. 331–49.

Hox, J. (1997), 'Multilevel Modeling: When and Why', in I. Balderjahn, R. Mathar and M. Schader (eds) *Classification, Data Analysis and Data Highways* (New York: Springer Verlag).

Hsiao, C. (1986), *Analysis of Panel Data* (Cambridge: Cambridge University Press).

Idachaba, F.S. (1994), 'Human Capital and African Agricultural Development', paper presented at the XXII International Conference of Agricultural Economists, 22–29 August (Harare, Zimbabwe).

IFAD (1992), Soil and Water Conservation in Sub-Saharan Africa. Towards Sustainable Production by the Rural Poor (Rome: IFAD).

INSAE (1979), 'Recensement Général de la Population et de l'Habitat, RGPH I' (Cotonou: Institut National de la Statistique et de l'Analyse Economique).

INSAE (1992), 'Recensement Général de la Population et de l'Habitat, RGPH II' (Cotonou: Institut National de la Statistique et de l'Analyse Economique).

INSAE (2000), 'Tableau de Bord Social' (Cotonou: Institut National de la Statistique et de l'Analyse Economique).

INSAE (2002), 'Recensement Général de la Population et de l'Habitat, RGPH III' (Cotonou: Institut National de la Statistique et de l'Analyse Economique).

INSEE (1961), 'Enquête Démographique au Dahomey' (Paris: Institut National de la Statistique et des Etudes Economiques).

Jacoby, H.G. (1993), 'Shadow Wages and Peasant Family Labour Supply: An Econometric Application to the Peruvian Sierra', *The Review of Economic Studies*, 60 (4) (October), pp. 903–21.

Jacoby, H. (2000), 'Access to Markets and the Benefits of Rural Roads', *The Economic Journal* 110 (July), pp. 713–37.

Jambiya, G. (1998), 'The Dynamics of Population, Land Scarcity, Agriculture and Non-agricultural Activities: West Usambara Mountains, Lushoto District, Tanzania' (Dar es Salaam/Leiden: DARE/ASC).

Jätzold, R. and H. Schmidt (1982), *Farm Management Handbook of Kenya* (Nairobi: Ministry of Agriculture).

Johnston, J. and J. Dinardo (1997), *Econometric Methods*, 4th edition (New York: McGraw-Hill International Editions).

Just, R.E. and Zilberman, D. (1983), 'Stochastic Structure, Farm Size and Technology Adoption in Developing Agriculture', *Oxford Economic Papers* 35, 307–28.

Kelly, V.A. (2006), 'Factors Affecting Demand for Fertilizer in Sub-Saharan Africa. Agriculture and Rural Development', discussion paper 23 (Washington, DC: The World Bank).

Klaasse Bos, A. and T. Dietz (1998), 'More People, Less Erosion: Environmental Recovery in Kenya. A Synthesis of Book Reviews and Resulting Questions', contribution to the opening workshop of the NWO programme on 'Agricultural Transition towards Sustainable Tropical Land Use', 25–28 March 1998 (Machakos, Kenya) (unpublished).

Knowler, D.J. (2004), 'The Economics of Soil Productivity: Local, National and Global Perspectives', *Land Degradation and Development* 15, pp. 543–61.

Knox, A., R. Meinzen-Dick and P. Hazell (2002), *Property Rights, Collective Action and Technologies for Natural Resource Management* (Washington, DC: IFPRI).

Kodjo, M., A. Kohoué, P. Gnagna, S. Gansou, T. N'Tcha (1995), 'La lutte anti-érosive et la restauration de la fertilité des sols dans le bassin versant de Koumagou, sous-préfecture de Boukombé: Etudes et analyse des différentes techniques mises en œuvre' (Cotonou: Projet de Gestion des Ressources Naturelles (PGRN)).

Kruseman, G. and J. Bade (1998), 'Agrarian Policies for Sustainable Land Use: Bio-economic Modelling to Assess the Effectiveness of Policy Instruments', *Agricultural Systems* 58, pp. 465–81.

Kurosaki, T. (1998), 'Risk and Household Behavior in Pakistan's Agriculture', Institute of Developing Economics, Occasional Papers Series No. 34.

Laman, M., H. Sandee, F. Zaal, H.A. Sidikou,.and E. Toe (1996), 'Combating Desertification – the Role of Incentives. A Desk Review of the Literature and

Practical Experiences with Incentives in Natural Resources Management in Sub-Saharan Africa, with an Emphasis on the Gender, Institutional and Policies Dimensions' (Rome: IFAD).

Lapar, M.L.A. and Pandey, S. (1999), 'Adoption of Soil Conservation: the Case of the Philippines Uplands', *Agricultural Economics* 21 (3), pp. 241–56.

Lee, R. (2003), 'The Demographic Transition: Three Centuries of Fundamental Change', *Journal of Economic Perspectives* 17 (4), 167–90.

Lindgren, Brit-Marie (1988), 'Machakos Report 1988: Economic Evaluation of a Soil Conservation Project in Machakos District, Kenya' (Nairobi: Ministry of Agriculture).

Lipton, M. (1979), 'Agricultural Risk, Rural Credit and Inefficiency of Inequality', in J.A. Roumasset, J.M. Boussard and I. Singh (eds) *Risk, Uncertainty and Agricultural Development* (Agricultural Development Centre).

Lipton, M. (1989), 'Responses to Rural Population Growth: Malthus and the Moderns', *Population and Development Review* 15, pp. 215–42.

Livi-Bacci, M. (1997), *A Concise History of World Population* (Oxford: Blackwell).

López, R. (1984), 'Estimating Labor Supply and Production Decisions of Self-Employed Farm Producers', *European Economic Review* 24, pp. 61–82.

López, R. (1998), 'Agricultural Intensification, Common Property Resources and the Farm-household', *Environmental and Resource Economics* 11 (3/4), pp. 443–58.

Maddala, G.S. (1983), *Limited-Dependent and Qualitative Variables in Econometrics* (Cambridge: Cambridge University Press).

Maddison, A. (1982), *Phases of Capitalist Development* (New York: Oxford University Press).

Mahmud, Y. (2004), 'Risk Time and Land Management under Market Imperfections: Applications to Ethiopia', PhD thesis (Department of Economics, Göteborg University).

Malthus, T. (1798), An Essay on the Principle of Population (London, printed for J. Johnson in St Paul's Church-yard).

Mangabat, C., D.J. Snelder, and W.T. de Groot (forthcoming), 'Motivating Tree Cultivation in NE Philippines: Financial Incentives, Technical Assistance and Good Communication', *Small-scale Forestry.*

Martin, J.Y. (1970), 'Les MATAKAM du Cameroun. Essai sur la dynamique d'une société préindustrielle' (Paris : ORSTOM).

Masipiqueña, A.B., M.D. Masipiqueña and W.T. de Groot (2008), 'Under-marketed and Over-regulated: The Case of Smallholder Tree Growing in Isabela, the Philippines', in D.J. Snelder and R.D. Lasco (eds) *Smallholder Tree Growing for Rural Development and Environmental Services* (Dordrecht: Springer Press), pp. 163–76.

Mazzucato, V. and D. Niemeijer (2000), 'The Cultural Economy of Soil and Water Conservation: Market Principles and Social Networks in Eastern Burkina Faso', *Development and Change* 31 (4).

Mbogoh, S. (2000), 'Makueni District Profile: Crop Production and Marketing', Drylands Research Working Paper 7 in the Kenya Series of Policy Requirements for Farmer Investment in Semi-Arid Africa (Somerset: Drylands Research).

Mbuvi, J.P. (2000), 'Makueni District Profile: Soil Fertility Management', Drylands Research Working Paper 6 in the Kenya Series of Policy Requirements for Farmer Investment in Semi-Arid Africa (Somerset: Drylands Research).

McFadden, D. (1983), 'Econometric Models of Probabilistic Choice', in C.F. Manski and D. McFadden (eds) *Structural Analysis of Discrete Data with Econometric Applications* (Cambridge: MIT Press).

Menezes, C., C. Geiss and J. Tessler (1980), 'Increasing Downside Risk', *American Economic Review* 70 (5), pp. 921–31.

Migot-Adhola, S.E., P.B. Hazell and F. Place (1990), 'Land Rights and Agricultural Productivity in Ghana, Kenya and Rwanda: A Synthesis of Findings' (Washington, DC: Agriculture and Rural Development Department, World Bank).

Minot, N. (1999), 'Effect of Transaction Costs on Supply Response and Marketed Surplus: Simulations using Non-separable Household Models', MTID Discussion Paper No. 36 (Washington, DC: IFPRI).

Miracle, M.P. (1967), *Agriculture in the Congo Basin* (Madison: University of Wisconsin Press).

Morgan, W.B. (1969), 'Peasant Agriculture in Tropical Africa', in M.F. Thomas and G.W. Whittington (eds) *Environment and Land-use in Africa* (London: Methuen).

Mortimore, M. (2001), 'Overcoming variability and productivity constraints in Sahelian agriculture', in T.A. Benjaminsen and C. Lund (eds) *Politics, Property and Production in the West African Sahel: Understanding Natural Resources Management* (Uppsala: Nordiska Afrikainstitutet), pp. 233–55.

Mortimore, M. (2002), 'Development and Change in Sahelian Dryland Agriculture', in D. Belshaw and I. Livingstone (eds) *Renewing Development in Sub-Saharan Africa: Policy, Performance and Prospects* (London: Routledge), pp. 135–52.

Mortimore, M. (2003), 'Long-term Change in African Drylands: Can Recent History Point Towards Development Pathways?' *Oxford Development Studies* 31 (4), pp. 503–18.

Mortimore, M. and W.M. Adams (1999), Working the Sahel: Environment and Society in Northern Nigeria (London: Routledge).

Mortimore, M. and W.M. Adams (2001), 'Farmer Adaptation, Change and "Crisis" in the Sahel', *Global Environmental Change* 11 (1), pp. 49–57.

Mortimore, M and M. Tiffen (1995), 'The Machakos Story', *Environment* 37 (7), pp. 5–7.

Mortimore, M., M. Tiffen, Y. Boubacar, J. Nelson (2001), 'Synthesis of long-term change in Maradi Department, Niger, 1960–2000', Drylands Research Working Paper 39 (Somerset: Drylands Research).

Mortimore, M. and M. Tiffen (1996), 'Population and Environment in Time Perspective: the Machakos Story', in T. Binns (ed.) *People and Environment in Africa* (Chichester: Wiley and Sons).

Mortimore, M. and M. Tiffen (2004), 'Introducing Research into Policy: Lessons from District Studies of Dryland Development in Sub-Saharan Africa', *Development Policy Review* 22 (3), pp. 259–86.

Mortimore, M. and M. Tiffen (1994), 'Population-growth and a Sustainable Environment', Environment 36 (8), pp. 10–13.

Moscardi, E. and A. de Janvry (1977), 'Attitudes toward Risk among Peasants: An Econometric Approach', *American Journal of Agricultural Economics* 59 (4), 710–16.

Mulder, I. (2000), 'Soil Degradation in Benin: Farmers' Perceptions and Responses', Ph.D. dissertation, Tinbergen Institute Research Series, No. 240 (VU University Amsterdam).

Murton, J. (1999), 'Population Growth and Poverty in Machakos District, Kenya', *Geographical Journal* 165 (1), pp. 37–46.

Mwakubo, S. (2002), 'Transaction Costs in Smallholder Agriculture: the Case of Soil Conservation in Kenya', Ph.D. thesis (Eldoret, Kenya: School of Environmental Studies, Moi University).

Myers, R.J. (1989), 'Econometric Testing for Risk Averse Behaviour in Agriculture', *Applied Economics* 21, pp. 541–55.

Nadel, S. (1942) *A Black Byzantium* (London: Oxford University Press).

Natta, J. (1999), 'Tradition et Développement: Occupation, Exploitation du Sol et Organisation Spatiale chez les Bètamaribè du Nord-Bénin', Mémoire de Maîtrise, Faculté des Lettres, Arts et Sciences Humaines (FLASH), Université Nationale du Bénin (UNB), Cotonou.

Nederveen Pieterse, J. (1998), 'My Paradigm or Yours? Alternative Development, Post-Development, Reflexive Development', *Development and Change* 29 (2), pp. 343–73.

Nelson, J. (2000), 'Makueni District Profile: Income Diversification and Farm investment, 1989–1999', Drylands Research Working Paper 10 in the Kenya Series of Policy Requirements for Farmer Investment in Semi-Arid Africa (Somerset: Drylands Research).

Netting, R. McC. (2003), Smallholders, Householders: Farm Families and the Ecology of Intensive, Sustainable Agriculture (Stanford, CA: Stanford University Press).

Norris, P.E. and S.S. Batie (1987), 'Virginia's Farmers' Soil Conservation Decisions: An Application of Tobit Analysis', *Southern Journal of Agricultural Economics* 19, pp. 79–90.

Nyang, F.O. (1999), 'Household Energy Demand and Environmental Management in Kenya', Ph.D. dissertation (University of Amsterdam).

Nzioka, C. (2000), 'Makueni District Profile: Human Resources Management, 1989–1998', Drylands Research Working Paper 9 in the Kenya Series of

Policy requirements for Farmer Investment in Semi-Arid Africa (Somerset: Drylands Research).

Okigbo, B.N. (1977), 'Farming Systems and Soil Erosion in West Africa', in D.J. Greenland and R. Lal (eds) *Soil Conservation and Management in the Humid Tropics* (Chichester: John Wiley).

Oostendorp, R.H. (1998), 'An Exploration of Alternative Modelling Approaches to the Malthus/Boserup Bifurcation', paper presented at the NWO Workshop on Agricultural Transition towards Sustainable Tropical Land Use (Machakos, Kenya), 25–28 March.

Overmars K.P. and P.H. Verburg (2006), 'Multilevel Modelling of Land Use from Field to Village Level in the Philippines', *Agricultural Systems*, 89 (2–3), pp. 435–56.

Overmars, K.P., W.T. de Groot and M.G.A. Huigen (2007), 'Comparing Inductive and Deductive Modelling of Land Use Decisions: Principles, a Model and an Illustration from the Philippines', *Human Ecology* 35, 439–52.

Pagiola, S. (1994), 'Soil Conservation in a Semi-Arid Region of Kenya: Rates of Return and Adoption by Farmers', in T. Naapier, S. Cambuni and S. El-Swoufy (eds), *Adopting Conservation on the Farm* (Ankeny, IA: Soil and Water Conservation Society).

Pagiola, S. (1996), 'Price Policy and Returns to Soil Conservation in Semi-arid Kenya', *Environmental and Resource Economics* 8 (3), pp. 255–71.

Pagiola, S., E. Lutz, and S. Reiche (1994), 'Soil Conservation in a Semi-Arid Region of Kenya: Rates of return and and adoption by farmers', in T.L. Napier, S.M. Camboni and S.A. El-Swaify (eds) *Adopting Conservation on the Farm, an International Perspective on the Socio-economics of Soil and Water Conservation* (Ankeny, IA: Soil and Water Conservation Society).

Pagiola, S., M. Mukumbu, M. Mutua and G. Okumu.(1990), 'Kitui District: The Effects of Environmental Degradation and Agricultural Policy on a Semi-Arid Area', Policy Analysis for Rural Development Working Paper No. 7 (Njoro: Egerton University).

Pender, J. (1998), 'Population Growth, Agricultural Intensification, Induced Innovation and Natural Resource Sustainability: An Application of Neoclassical Growth Theory', *Agricultural Economics* 19, 99–112.

Pender, J. (1999), 'Rural Population Growth, Agricultural Change, and Natural Resource Management in Developing Countries: a Review of Hypotheses and Some Evidence from Honduras', EPTD Discussion Paper 48 (Washington DC: IFPRI).

Pender, J.L., Jagger, P., Nkonya, E. and Sserunkuuma, D. (2004), 'Development Pathways and Land Management in Uganda', *World Development* 32 (5), pp. 767–92.

Pender, J. and J. Kerr (1996), 'Determinants of Farmers' Indigenous Soil and Water Conservation Investments in India's Semi-arid Tropics', EPTD Discussion Paper No. 17 (Environment and Production Technology Division, International Food Policy and Research Institute).

Pender, J.L. and J.M. Kerr (1998), 'Determinants of Farmers' Indigenous Soil and Water Conservation Investments in Semi-arid India', *Agricultural Economics* 19, pp. 113–25.

Petrick, M. (2004), 'A Microeconometric Analysis of Credit Rationing in the Polish Farm Sector', *European Review of Agricultural Economics* 31 (1), pp. 77–101.

Pingali, P. and H.P. Binswanger (1984), 'Population Density and Agricultural Intensification: A Study of the Evolution of Technologies in Tropical Agriculture', Agricultural and Rural Development Department, Research Unit, report N° ARU 22 (Washington, DC: World Bank).

Pingali, P. and H.P. Binswanger (1988), 'Population Density and Farming Systems: The Changing Locus of Innovations and Technical Change'. In R. Lee, W.B. Arthur, A.C. Kelley, G. Rodgers and T.N. Srinivasan (eds) *Population, Food and Rural Development* (Oxford: Clarendon Press).

Pingali, P.L., P.F. Moya and L.E. Velasco (1990), 'The Post Green Revolution Blues in Asian Rice Production: The Diminished Gap between Experiment Station and Farmer Yields', Social Sciences Division Paper N° 90-01 (Los Bãnos, Laguna, Philippines: International Rice Research Institute (IRRI)).

Place, F. and P. Hazell.(1993), 'Productivity Effects of Indigenous Land Tenure Systems in Sub-Saharan Africa',.*American Journal of Agricultural Economics* 75, pp. 10–19.

Platteau, J.-Ph. (2000), *Institutions, Social Norms and Economic Development* (Amsterdam: Harwood Academic Publishers).

Pomp, M. and Burger, K. (1995), 'Innovation and Imitation: Adoption of Cocoa by Indonesian Smallholders', *World Development* 23 (3), pp. 423–31.

Preston, D., M. Macklin and J. Walburton (1997), 'Fewer People, Less Erosion: the Twentieth Century in Southern Bolivia', *The Geographical Journal* 163 (July), pp. 198–205.

Pudasaini, S.P. (1983), 'The Effect of Education in Agriculture: Evidence from Nepal', *American Journal of Agricultural Economics* 65 (August), pp. 509–15.

Rahman, S.A., de Groot, W.T. and Snelder, D.J. (2008), 'Exploring the Agroforestry Adoption gap: the Financial and Social Economics of Agroforestry by Smallholders in Bangladesh', in D.J. Snelder and R.D. Lasco (eds) *Smallholder Tree Growing for Rural Development and Environmental Services* (Dordrecht: Springer Press), pp. 227–44.

Randall, A. (1981), 'Property Entitlement and Pricing Policy for a Maturing Water Economy', *Australian Journal of Agricultural Economics* 25 (3), pp. 195–220.

Reardon, T. and S.A. Vosti (1992), 'Issues in the Analysis of the Effects of Policy on Conservation and Productivity at the Household Level in Developing Countries', *Quarterly Journal of International Agriculture* 31 (4), pp. 380–96.

Reardon, T. and S.A. Vosti (1995), 'Links Between Rural Poverty and the Environment in Developing Countries: Asset Categories and Investment Poverty', *World Development* 23 (9), pp. 1495–506.

Reij, C., I. Scoones and C. Toulmin (1996a) *Techniques traditionnelles de conservation des eaux et des sols en Afrique* (Paris: Ed. Karthala; Amsterdam: CDCS and Wageningen: CTA).

Reij, C., Scoones, I. and Toulmin, C. (1996b), *Sustaining the Soil. Indigenous Soil and Water Conservation in Africa* (London: Earthscan).

Reij, C. and Waters-Bayer, A. (2001), Farmer Innovation in Africa. A Source of Inspiration for Agricultural Development (London: Earthscan Publications).

Ribot, J.C. (1999), 'Decentralisation, Participation and Accountability in Sahelian Forestry: Legal Instruments of Political-administrative Control', *Africa* 69 (1), pp. 23–65.

Richards, A.I. (1939) *Land, Labour and Diet in Northern Rhodesia: An Economic Study of the Bemba Tribe* (London: Oxford University Press).

Richards, P., A.O. Phillips and L.J. Slikkerveer (1989), *Indigenous Knowledge Systems for Agriculture and Rural Development: the Cikard Inaugural Lectures* (Ames, IO: Iowa State University).

Rocheleau, D.E., P.E. Sternberg and P.A. Benjamin (1995), 'Environment, Development Crisis and Crusade, Ukambani Kenya, 1890–1990', *World Development* 23 (6), pp. 1037–51.

Romero, M.R. and W.T. de Groot (2008), 'Income Security, Knowledge and Village Factors: Explaining Transition towards Sustainable Land Use in a Tropical Forest Fringe', in R.B. Dellink and A. Ruijs (eds), *Economics of Poverty, Environment and Natural Resource Use* (Dordrecht: Springer) pp. 157–84.

Romero, M.R. (2006), 'Investing in the Land; Agricultural Transition towards Sustainable Land Use in the Philippines Forest Fringe', PhD thesis (Leiden University).

Ruben, R., G. Kruseman, H. Hengsdijk and A. Kuyvenhoven (1996), 'The Impact of Agrarian Policies on Sustainable Land Use', mimeo (Agricultural University, Wageningen).

Scherr, S.J. and P. Hazell (1994), 'Sustainable Agricultural Development Strategies in Fragile Lands', EPTD Discussions paper No.1 (Washington, DC: IFPRI).

Scoones, I. (1999), 'New Ecology and the Social Sciences. What Prospects for a Fruitful Engagement?', *Annual Review of Anthropology* 28, pp. 479–507.

Seignobos, C. and O. Iyébi-Mandjek (2000), *Atlas de la Province Extrème-Nord, Cameroun* (Paris: IRD Editions).

Sellen, D., G. Argwings-Kodhek, A. Chomba, F. Karin, and M. Mutua (1990), 'Agriculture in Nyeri: Farm Income, Efficiency, and Agricultural Policy', Policy Analysis for Rural Development Working Paper No. 11 (Njoro: Egerton University).

Sengalawe, Z.M. (1998), 'Household Adoption Behaviour and Agricultural Sustainability in the North-Eastern Mountains of Tanzania', PhD thesis (Wageningen University).

Shiferaw, B. and S. Holden (1996), 'Resource Degradation and Adoption of Land Conservation Technologies in the Ethiopian Highlands: A Study in Andit Tid, North Shewa', Discussion Paper D-31/1996 (Ås: Agricultural University of Norway).

Shiferaw, B. and S. Holden (1998), 'Investment in Soil and Water Conservation in the Ethiopian Highlands: Does it Pay Small Farmers?', Paper No. D-32/1998, Department of Economics and Social Sciences (Ås: Agricultural University of Norway).

Shiferaw, B. and S. Holden (2000), 'Policy Instruments for Sustainable Land Management: the Case of Highland Smallholders in Ethiopia', *Agricultural Economics* 22 (3), pp. 217–32.

Shively, G. (1996), 'Assets, Attitudes, Beliefs, and Behaviors: Explaining Patterns of Soil Conservation Adoption among Low-income Farmers', Working Paper No. 19 (Los Baños, Laguna: SEARCA).

Shultz, T. (1964), *Transforming Traditional Agriculture* (New Haven: Yale University Press).

Siedenburg, J. (2006), 'The Machakos Case Study: Solid Outcomes, Unhelpful Hyperbole', *Development Policy Review* 24 (1), pp. 75–85.

Sinden, J.A. and D.A. King (1988), 'Land Condition, Crop Productivity, and the Adoption of Soil Conservation Measures', paper presented at the Australian Agricultural Economics Society Conference, Melbourne, Australia.

Singh, L.P., B. Singh and H.S. Bal (1987), 'Indiscriminate Fertilizer Use *vis-à-vis* Ground Water Pollution in Central Punjab', *Indian Journal of Agricultural Economics* 42 (3), pp. 404–409.

Smith, J. and A.D. Barau (1994), 'The Role of Technology in Agricultural Intensification: the Evolution of Maize Production in the Northern Guinea Savanna of Nigeria', *Economic Development and Cultural Change* 42 (3), pp. 537–53.

Solow, R.M. (1974), 'The Economic of Resources or the Resources of Economics', *American Economic Review* 64 (2) pp. 1–14.

Spencer, D. (1994), 'Infrastructure and Technological Constraints to Agricultural Development in the Humid and Subhumid Tropics of Africa', EPTD Discussions Paper No. 3 (Washington, DC: IFPRI).

Stone, M.P., G.D. Stone and R.McC. Netting (1995), 'The Sexual Division of Labor in Kofyar Agriculture', *American Ethnologist* 22 (1), pp. 165–86.

Subba Rao, D.V., K.R. Chowdry and G.G. Venkata Reddy (1987), 'Degradation of Agro-ecosystem – an Exploratory Study on Cotton Farming', *Indian Journal of Agricultural Economics* 42 (3), pp. 410–15.

Swinton, S.M., G. Escobar and T, Reardon (2003), 'Poverty and Environment in Latin America: Concepts, Evidence and Policy Implications', *World Development* 31(11), pp. 1865–72.

Taylor, J. and I. Adelman (1996), *Village Economies: The Design, Estimation, and Use of Village Wide Economic Models* (Cambridge: Cambridge University Press).

Templeton, S. (1994), 'Microeconomic Analysis of Land Management: the Control of Soil Erosion and an Empirical Analysis of Non-paddy Terracing in the Philippines', Ph.D. dissertation (Berkeley: University of California).

Templeton, S. and S. Scherr (1997), 'Population Pressure and the Microeconomy of Land Management in Hills and Mountains of Developing Countries', EPTD Discussion Paper (Washington, DC: IFPRI).

Tiffen, M. (1991), 'Environmental Change and Dryland Management in Machakos District, Kenya 1930–1990. Production Profile', working paper (London: Overseas Development Institute).

Tiffen, M. (1992), 'Environment, Population Growth and Productivity in Kenya. A Case Study of Machakos District', paper presented for the annual conference of ESRC Development Economics Group, University of Leicester.

Tiffen, M. (1995), 'Population Density, Economic Growth and Societies in Transition. Boserup Reconsidered in a Kenyan Case Study', *Development and Change* 26 (1), pp. 31–65.

Tiffen, M. (2002), 'The Evolution of Agroecological Methods and the Influence of Markets: Case Studies from Kenya and Nigeria', in N. Uphoff (ed.) *Agroecological Innovations: Increasing Food Production with Participatory Development* (London: Earthscan), pp. 95–108.

Tiffen, M. (2003), 'Transition in Sub-Saharan Africa: Agriculture, Urbanization and Income Growth', *World Development* 31 (8), pp. 1343–66.

Tiffen, M. (2006), 'Urbanization: Impacts on the Evolution of "Mixed Farming" Systems in Sub-Saharan Africa', *Experimental Agriculture* 42 (3), pp. 259–87.

Tiffen, M. and Mortimore, M. (1993), 'Population Growth and Natural Resource Use: Do We Need to Dispair for Africa?' *Outlook for Agriculture* 22 (4), pp. 241–9.

Tiffen, M. and M. Mortimore (1994), 'Malthus Controverted – the Role of Capital and Technology in Growth and Environment Recovery in Kenya', *World Development* 22 (7), pp. 997–1010.

Tiffen, M. and M. Mortimore (2002), 'Questioning Desertification in Dryland Sub-Saharan Africa', *Natural Resources Forum* 26 (3), pp. 218–33.

Tiffen, M. and M. Mortimore (2006), 'Response: Forwards to New Challenges, Not Backwards to Prescription', *Development Policy Review* 24 (1), pp. 87–104.

Tiffen, M., M. Mortimore and F. Gichuki (1994), *More People, Less Erosion: Environmental Recovery in Kenya* (London: Overseas Development Institute/ Wiley; Nairobi: ACTS Press).

Tiffen, M., R. Purcel, F. Gichuki, C. Gachene and J. Gatheru (1996) 'National Soil and Water Conservation Programme, SIDA Evaluation 96/25' (Stockholm: Department for Natural Resources and the Environment, SIDA).

Timmermans, D.S. and W.T. de Groot (2002),'"One Tree is Better than Three Cows", Motivations to Invest in Fruit Orchards, North Cameroon', in M. Ali, P.E. Loth, H. Bauer and H.H. de Iongh (eds), *Management of Fragile*

*Ecosystems in North Cameroon* (Maroua: Centre d'Études de l'Environnment et Développement), pp. 141–58.

Turner, B.L., II, G. Hyden and R.W. Kates (eds) (1993), *Population Growth and Agricultural Change in Africa* (Gainesville, FL: University Press Florida).

Uphoff, N., M.J. Esman and A. Krishna (1998), *Reasons for Success: Learning from Instructive Experiences in Rural Development* (West Hartford, CN: Kumarian Press).

Uphoff, N. and C.M. Wijayaratna (2000), 'Demonstrated Benefits from Social Capital: The Productivity of Farmer Organizations', *World Development* 28 (11), pp. 1975–890.

Upton, M. (1994), book review, *Development Policy Review* 12, pp. 328–34.

Van den Berg, M. (2002), 'Do Public Works Decrease Farmers' Soil Degradation? Labour Income and the Use of Fertilizers in India's Semi-arid Tropic', *Environment and Development Economics* 7, pp. 487–506.

Van den Eeden, P. and H.J.M. Hüttner (1982), 'The Multilevel Approach', *Current Sociology* 1982 (30), pp. 26–38.

Wachter, D. (1992), 'Land Titling for Land Conservation in Developing Countries', Environment Department Divisional Working Paper 1992–28 (Washington, DC: World Bank).

Wadsworth, R. and Swetnam, R. (1998), 'Modelling the Impact of Climate Warming at the Landscape Scale: will Bench Terraces become Economically and Ecologically Viable Structures under Changed Climates?', *Agriculture, Ecosystems and Environment* 68 (1–2), pp. 27–39.

Walker, D.J. (1982), 'A Damage Function to Evaluate Erosion Control Economics', *American Journal of Agricultural Economics* 64 (4, November), pp. 145–55.

Wassouni (2006), 'Bushland in the Mindif Region, Cameroon: Functions, Decline, Context and Prospects', PhD dissertation (Leiden University; available in the Leiden repository: https://openaccess.lcidenuniv.nl).

Welch, F. (1978), 'The Role of Investments in Human Capital in Agriculture', in T.W. Schult (ed.) *Distortions of Agricultural Incentives* (Indiana University Press) pp. 259–80.

Wiggins, S. (2000), 'Interpreting Changes from the 1970s to the 1990s in African Agriculture through Village Studies', *World Development* 28 (4), pp. 631–62.

Winter-Nelson, A., F. Karin, and M. Nyambu (1990), 'Agriculture in Nakuru District: Strategies for Growth', Policy Analysis for Rural Development Working Paper No. 4 (Njoro: Egerton University).

Wolgin, J.M. (1975), 'Resource Allocation and Risk: A Case of Small Holder Agriculture in Kenya', *American Journal of Agricultural Economics* 57 (4), pp. 622–30.

World Bank (1994), 'The World Bank and Participation', Report of the Learning Group on Participatory Development. April (Washington, DC: The World Bank).

Yesuf, M. (2004), 'A Dynamic Economic Model of Soil Conservation with Imperfect Market Institutions', doctoral dissertation, University of Gothenburg.

Zaal, F. (1999), 'The Agricultural Transition towards Sustainable Tropical Land Use', presentation of preliminary results in Kenya; paper, presented at the 'NWO Programme on Environment and the Economy' workshop, Oibibio Amsterdam, 2 June.

Zaal, F., M. Laman and C.M. Sourang (1998), 'Resource Conservation or short-term food needs? Designing incentives for natural resource management', *IIED Issues paper* no 77, April 1998. London: IIED.

Zaal, F. and Oostendorp, R. (2002), 'Explaining a Miracle: Intensification and the Transition Towards Sustainable Small-scale Agriculture in Dryland Machakos and Kitui Districts, Kenya', *World Development* 30 (7), July, pp. 1271–87.

Zuiderwijk, A.B. (1995), 'World Market Crops on Vulnerable Soils. The Case of Cotton Cultivation along the Mandara Mountains of North Cameroon', paper presented at the Agricultural Questions Conference, Wageningen.

Zuiderwijk, A.B. (1997), 'Much Circumstantial Evidence, Few Eye-Witnesses', paper presented at the Opening Workshop of the NWO Program 'The Agricultural Transition Towards Sustainable Tropical Land Use', Machakos, Kenya, 25–28 March 1998.

Zuiderwijk, A.B. (1998), 'Farming Gently, Farming Fast', PhD dissertation (Leiden University, CML, Leiden).

Zuiderwijk, A.B. and J. Schaafsma (1997), 'Male Outmigration, Changing Women's Roles and Consequences for Environmental Management; the Case of the Mafa in North Cameroon', in M. de Bruyn (ed.), *Gender and Land Use; Diversity in Environmental Practices* (Amsterdam: Thela Publishers).

# Index

Note: Bold page numbers indicate tables and figures. Numbers in brackets preceded by *n* are footnote numbers.